Marco Calisto

Influence of energetic particles on atmospheric chemistry and climate

Marco Calisto

Influence of energetic particles on atmospheric chemistry and climate

Galactic cosmic rays, solar protons and energetic particles and their influence on the Earth's atmosphere

Südwestdeutscher Verlag für Hochschulschriften

Impressum/Imprint (nur für Deutschland/only for Germany)
Bibliografische Information der Deutschen Nationalbibliothek: Die Deutsche Nationalbibliothek verzeichnet diese Publikation in der Deutschen Nationalbibliografie; detaillierte bibliografische Daten sind im Internet über http://dnb.d-nb.de abrufbar.

Alle in diesem Buch genannten Marken und Produktnamen unterliegen warenzeichen-, marken- oder patentrechtlichem Schutz bzw. sind Warenzeichen oder eingetragene Warenzeichen der jeweiligen Inhaber. Die Wiedergabe von Marken, Produktnamen, Gebrauchsnamen, Handelsnamen, Warenbezeichnungen u.s.w. in diesem Werk berechtigt auch ohne besondere Kennzeichnung nicht zu der Annahme, dass solche Namen im Sinne der Warenzeichen- und Markenschutzgesetzgebung als frei zu betrachten wären und daher von jedermann benutzt werden dürften.

Verlag: Südwestdeutscher Verlag für Hochschulschriften GmbH & Co. KG
Dudweiler Landstr. 99, 66123 Saarbrücken, Deutschland
Telefon +49 681 37 20 271-1, Telefax +49 681 37 20 271-0
Email: info@svh-verlag.de

Zugl.: Zürich, ETH, Diss, 2011

Herstellung in Deutschland:
Schaltungsdienst Lange o.H.G., Berlin
Books on Demand GmbH, Norderstedt
Reha GmbH, Saarbrücken
Amazon Distribution GmbH, Leipzig
ISBN: 978-3-8381-2850-4

Imprint (only for USA, GB)
Bibliographic information published by the Deutsche Nationalbibliothek: The Deutsche Nationalbibliothek lists this publication in the Deutsche Nationalbibliografie; detailed bibliographic data are available in the Internet at http://dnb.d-nb.de.

Any brand names and product names mentioned in this book are subject to trademark, brand or patent protection and are trademarks or registered trademarks of their respective holders. The use of brand names, product names, common names, trade names, product descriptions etc. even without a particular marking in this works is in no way to be construed to mean that such names may be regarded as unrestricted in respect of trademark and brand protection legislation and could thus be used by anyone.

Publisher: Südwestdeutscher Verlag für Hochschulschriften GmbH & Co. KG
Dudweiler Landstr. 99, 66123 Saarbrücken, Germany
Phone +49 681 37 20 271-1, Fax +49 681 37 20 271-0
Email: info@svh-verlag.de

Printed in the U.S.A.
Printed in the U.K. by (see last page)
ISBN: 978-3-8381-2850-4

Copyright © 2011 by the author and Südwestdeutscher Verlag für Hochschulschriften GmbH & Co. KG and licensors
All rights reserved. Saarbrücken 2011

DISS. ETH NO. 19252

Influence of energetic particle precipitation on atmospheric chemistry and climate

A dissertation submitted to the

ETH ZURICH

for the degree of

Doctor of Sciences

presented by

MARCO CALISTO

Dipl. Natw. ETH, Switzerland

born 15 February 1973

citizen of Illnau-Effretikon (Zurich)

accepted on the recommendation of

Prof. Dr. Thomas Peter, examiner
Dr. Eugene Rozanov, co-examiner
Dr. Thomas Reddmann, co-examiner

2011

Contents

Abstract	7
Zusammenfassung	11
1 Introduction	**1**
1.1 Motivation	1
1.2 Objectives and outline	3
2 Energetic particles	**5**
2.1 Galactic Cosmic Rays	7
2.1.1 Origin and characteristics	7
2.1.2 Interaction with Earth's atmosphere	8
2.2 Solar Proton Events	9
2.2.1 Origin and characteristics	9
2.2.2 Interaction with Earth's atmosphere	11
2.3 Energetic Electron Precipitation	12
2.3.1 Origin and characteristics	12
2.3.2 Interaction with Earth's atmosphere	14
3 Chemistry and dynamics	**17**
3.1 The role of Ozone and its chemistry	17
3.1.1 The distribution of ozone in the atmosphere	19
3.2 NOx chemistry	21
3.3 HOx chemistry	23

3.4	Halogen chemistry	25
	3.4.1 Chlorine	25
	3.4.2 Bromine	26
3.5	Heterogeneous chemistry	26
3.6	Dynamics and transport	27
	3.6.1 The Polar Vortex	28
	3.6.2 The Brewer Dobson Circulation	29
	3.6.3 Waves	29
4	**Model description, parameterizations and experimental setup**	**31**
4.1	Model description	31
4.2	Parameterizations	32
	4.2.1 GCR runs	33
	4.2.2 SPE runs	38
	4.2.3 EEP runs	41
	4.2.4 Coupled run	43
4.3	Experimental setup	44
5	**GCR results**	**47**
6	**SPE results**	**71**
7	**GCR, SPE and LEE simultaneously**	**89**
7.1	Introduction	89
7.2	Description of the Model and experimental setup	92
	7.2.1 Chemistry-Climate Modeling	92
	7.2.2 Energetic Particles Induced Modeling	92
7.3	Results	94
7.4	Conclusion	104

Contents

8 Conclusion and Outlook — 105
    8.1 Conclusion — 105
    8.2 Outlook — 107

A HEPPA intercomparison — 109
    A.1 NOx comparison — 110
    A.2 HNO_3 comparison — 112
    A.3 Ozone comparison — 114

List of Figures — 115

List of Tables — 123

Bibliography — 125

Curriculum Vitae — 127

Acknowledgements — 129

CONTENTS

Abstract

Solar activity has a large influence on the Earth and upon life on Earth, not only the energy that is sent to us in the form of light but also energetic particles, that are sent to our planet by the Sun, and other energetic particles (Galactic cosmic rays and energetic electrons) that are modulated by the Sun's activity.

During solar flares (large explosions on the Sun's surface), solar protons are thrown out and are guided by the interplanetary magnetic field (IMF) through space. The intensity of Galactic cosmic rays (GCRs), which consist mostly of protons coming from outer space, is coupled to the Sun's activity; the input of GCRs to the atmosphere is largest when solar activity is low and weakest when it is high. The reason for this is that during times when the Sun is active, the IMF is stronger and therefore the GCRs are deflected, only highly energetic particles are able to penetrate through the IMF during these times. The energetic electrons that are trapped in the Van Allen belts are pushed towards the Earth during times when the magnetic activity of the Sun is strong (energetic electron precipitation EEP).

If these particles are energetic enough they can enter deeply into the Earth's atmospere at the polar cap regions (geomagnetic latitude $> 60^o$), and then react with the chemical species in the atmosphere. The Galactic cosmic rays have energies that are high enough to enter the atmosphere at all latitudes, i.e. they are not restricted to the polar regions. The high kinetic energy of the particles leads to the ionization of neutral constituents of the Earth's atmosphere and thereby to additional production of HO_x and NO_x which can accelerate the ozone depletion cycles in the middle atmosphere. The ozone loss, in turn, can induce changes in the zonal wind, which can have an impact on climate.

The goal of this thesis is to investigate the effects of particle precipitation on atmospheric chemistry and dynamics caused by the ionization of the Earth's atmosphere through the above mentioned particles. To this end, the 3-D chemistry-climate model SOCOL v2.0 (**SO**lar **C**limate **O**zone **L**inks), which covers the atmosphere from the surface up to 80 km, has been employed,

Abstract

using parameterizations for the different energetic particles that had to be found in the literature and/or developed.

First, an extensive validation has been performed with the model runs that have been executed with the influence of the GCRs. The runs covered several solar cycles, i.e. they started 1976 and ended in 2002 because the intensity of the Galactic cosmic rays are inversely related to the solar activity, i.e. during solarmin the flux of the GCRs is highest and vice versa. Overall, it can be concluded that the influence of the GCRs on the atmospheric chemistry and dynamics cannot be neglected when looking at the troposphere and the UTLS-region.

The analysis of the results show that the GCRs, which are the particles with the highest energies (up to a few GeV), penetrate down to the troposphere causing chemical changes in this height-range. In fact, an increase in NO_x is visible almost throughout all latitudes and altitudes. The most pronounced increase of up to 24 % is ocurring in the pristine Southern Hemispheric (SH) upper troposphere. This increase in NO_x induces tropospheric ozone production. In contrast, the northern hemisphere shows a decrease in ozone of more than 3 % in the lower stratosphere which causes a change in the meridional temperature distribution. This leads to an acceleration of the polar night jet and a strenghtening of the northern polar vortex, extending from the middle stratosphere all the way to the ground. Remarkably, the surface air temperature over the eastern part of Europe and Russia and over Greenland are subject to significant changes, with warming in Eurasia and cooling in Greenland due to dynamical reasons triggered by the ozone decrease in the northern hemisphere (Thompson & Wallace, 1998).

Second, for the solar protons the well known October/November 2003 event and an event similiar to the Carrington event from September 1859 have been chosen, for which several ensemble runs have been performed and validated against satellite data and previously published results. The runs show that these solar proton events (SPEs) had a statistically significant impact on NO_x, HO_x, ozone and the zonal wind.

Because solar proton events last for just a few days, their impact on the atmospheric chemistry is often most pronounced shortly after the event started. The analysis showed that the increase for NO_x of up to 300 ppb and HO_x of approximately 8 ppb occurs in the southern hemispheric polar region shortly after the event happened. The decrease for ozone of more than 60 % in the same region is a consequence to the increase of NO_x and HO_x. Because solar protons are less energetic than GCRs (up to 500 MeV) they cannot penetrate down to the troposphere. The most intense interaction is at heights from 70

Abstract

km down to 40 km, even though very rarely the solar protons are energetic enough to reach the surface. The increase of NO_x and HO_x are most pronounced in polar night regions. Depending on how strong the SPE was, the impact on NO_x and ozone remains significant for several weeks.

For runs concerning energetic electrons, time periods have been taken when the Sun is in a magnetically active phase and also when the magnetic activity of the Sun is low to be able to compare the intensity of the energetic electrons. During the magnetically active phase of the Sun, the precipitation of energetic electrons to the Earth's atmosphere out of the Van Allen belt is strongest. The model runs with the low and high energetic electrons reveal statistically significant changes for NO_x, HO_x and ozone from the mesosphere down to the lower stratosphere.

Energetic electrons are the particles with the lowest energy. They can have energies up to 10 MeV. Due to this fact, the height where the electrons ionize the air varies from 110 km down to 70 km. Like the GCRs and the SPEs, additional NO_x and HO_x will be produced and can be transported down with time which leads to a decrease of ozone that is most pronounced during polar winter when the vortex hinders the midlatitude ozonrich air to enter in the polar region.

Abstract

Zusammenfassung

Die Aktivität der Sonne hat einen grosse Einfluss auf unsere Erde und das Leben darauf. Jedoch nicht nur mit der Energie, die durch die Sonne in Form von Licht zu uns transportiert wird, sondern auch durch die energetischen Partikel, welche von der Sonne auf unseren Planeten kommen, sowie anderen energetischen Partikeln, z.b. den Galaktischen Kosmischen Strahlen (GCRs) und den energetischen Elektronen (EEP), welche durch die Aktivität der Sonne moduliert werden.

Solare Protonen (SPEs) werden während starken Eruptionen auf der Sonnenoberfläche von der Sonne weggeschleudert und durch das Interplanetare Magnetfeld (IMF) weggeleitet. Die intensität der GCRs, welche zum grössten Teil aus Protonen bestehen die vom inneren der Galaxis kommen, sind nur indirekt von der Sonne beeinflusst, das heisst während des Solaren Maximums ist die Intensität der GCRs minimal und umgekehrt. Der Grund ist, dass das IMF während dieser Zeit durch die Sonne moduliert wird und stärker ist. Nur die hochenergetischsten Partikel werden nicht reflektiert und können daher bis zur Erde vordringen. Die energetischen Elektronen sind normalerweise in dem Van Allen Gürtel gefangen und werden während Phasen, in der die magnetische Aktivität der Sonne gross ist, näher zur Erdoberfläche gedrückt.

Vorausgesetzt diese Partikel besitzen genug Energie, können sie in die Erdatmosphäre auf der Höhe der Polaren Regionen (geomagnetische Latitude $> 60^o$) eindringen. Die GCRs sind so hochenergetisch, dass sie überall auf der Erde eindringen können, das heisst, sie sind nicht auf die Polaren Regionen beschränkt. Diese durch die Partikel hervorgerufene zusätzliche Energie, führt zur Ionisation der Atmosphäre, was zu zusätzlichem HO_x und NO_x führen kann. Dieses HO_x und NO_x aktivieren den Ozonzerstörungs Zyklus in der Atmosphäre, wobei die Lebensdauer des HO_x Zyklus geringer ist als der des NO_x Zyklus. Die durch die Partikel hervorgerufene Zerstörung des Ozons, kann zu Änderungen in den Zonalen Winden führen welche dann Auswirkungen auf das Klima haben.

Die hier vorliegende Arbeit hat zum Ziel, die Änderungen in der Atmosphärenchemie und der Dynamik, hervorgerufen durch die energetischen Par-

Zusammenfassung

tikel, zu untersuchen. Hierfür wird das 3-D Klima-Chemie-Model SOCOL v2.0 (**SO**lar **C**limate and **O**zone **L**inks) benutzt welches die Atmosphäre vom Boden bis in eine Höhe von 80 km abdeckt. Um das Ziel erreichen zu können, mussten Parameterisierungen für die GCRs, SPEs und EEPs entwickelt und/oder gefunden werden. Als nächstes wurden verschiedene Modellruns mit den Energetischen Partikeln durchgeführt.

Als erstes wurden Modelläufe mit dem Einfluss der GCRs ausgeführt und validiert. Die Modelläufe dauerten mehrere Solare Zyklen, d.h. sie starteten jeweils 1976 und endeten 2002 da die Intensität der Galaktischen Kosmischen Strahlung mit dem Sonnenzyklus variiert. Im Grossen und Ganzen konnte gefolgert werden, dass der Einfluss der GCRs auf die Atmosphärenchemie und die Dynamik wichtig ist wenn man die Troposphäre und die UTLS-region untersucht.

Danach wurde für die SPEs der Event welcher im Oktober/November 2003 stattfand, und ein Event der ähnlich dem Carrington event vom September 1859 ist, ausgesucht und mit Satellitendaten und schon publizierten Resultaten verglichen. Es konnte gezeigt werden dass diese SPEs einen statistisch signifikante Einfluss auf NO_x, HO_x, Ozon und den Zonalen Wind haben.

Für die Modelläufe mit dem Einfluss der energetischen Elektronen wurden Perioden gewählt, in denen die Magnetische Aktivität der Sonne stark, wie auch schwach ist, damit untersucht werden kann, wie stark diese Partikel die Atmosphäre beeinflussen. Während der Zeit, in der die Sonne magnetisch aktiv ist, werden am meisten energetische Elektronen vom Van Allen Gürtel gegen die Erdatmosphähre gedrückt. Die Modelläufe haben gezeigt, dass die energetischen Elektronen von der Mesosphäre bis zur unteren Stratosphäre einen statistisch signifikanten Einfluss auf NO_x, HO_x, Ozon haben.

Die Analyse der Resultate hat gezeigt, dass die GCRs, welche die energetischsten Partikel sind (bis mehrere GeV), bis zur Troposphäre vordringen können. Das zusätzlich produzierte NO_x ist fast über alle Latituden und Höhen sichtbar. Die grösste Änderung von bis zu 24 % ist in der Südhemisphärischen Troposphäre zu sehen. Diese Zunahme an NO_x initiiert den Ozonproduktionszyklus der Troposphäre. Im Gegenzug sieht man in der Nordhemisphäre eine Reduktion von mehr als 3 % im Ozon, welches eine Änderung des Temperaturregimes in Gang bringt. Diese Temperaturänderung kann den Zonalen Wind ändern, welcher dann eine Änderung der Oberflächentemperatur in den östlichen Teilen von Europa, Russland und über Grönland mit sich zieht.

Da die SPEs nur wenige Tage andauern ist ihr Einfluss normalerweise am stärksten kurz nachdem der Event gestartet ist. Die Analyse hat gezeigt,

Zusammenfassung

dass es eine Zunahme von bis zu 300 ppb bei den NO_x Spezies geben kann, sowie ungefähr 8 ppb bei den HO_x Spezies. Die Abnahme von mehr als 60 % beim Ozon ist eine Konsequenz der Zunahme der NO_x und HO_x Spezies. Da die Solaren Protonen weniger hochenergetisch sind als die GCRs (bis 500 Mev), reichen sie nicht bis zur Troposphäre. Die meiste Energie geht durch ionisation in Höhen von 40 km bis 70 km verloren, obwohl es doch vorkommt, dass solare Protonen bis zur Erdoberflche eindringen können. Der zusätzliche Input von HO_x und NO_x ist in Gebieten in denen noch polare Nacht herrscht am Besten zu sehen. Das NO_x kann durch Transport bis in die untere Stratosphäre gebracht werden, wo es dann auch ozonzerstörend reagieren kann. Der Einfluss der solaren Protonen kann noch während Wochen sichtbar sein, je nachdem wie stark der Event war.

Die energetischen Elektronen sind die niederenergetischsten Partikel (bis 10 MeV). Aus diesem Grund ionisieren die Elektronen die Erdatmosphäre in einer Höhe von 110 km bis zu 70 km. Genau wie die GCRs oder die SPEs wird auch hier zusätzliches NO_x und HO_x produziert welches durch Transport in die Stratosphäre gelangen kann und dadurch ozonzerstörend wirken kann. Während der polaren Nacht findet diese Zerstörung des ozons am stärksten statt, da der Polare Vortex die ozonhaltige Luft aus den mittleren Breiten daran hindert in die polaren Regionen vorzudringen.

Zusammenfassung

Chapter 1

Introduction

1.1 Motivation

The relationship between the Earth and the Sun/outer space has been an interest for researchers for many years. They all were curious whether and how the Earth's atmosphere interacts with the variable Sun and outer space. It was already known that different natural phenomena are able to change the atmospheric chemistry, e.g. volcanic eruptions, El Nino, La Nina, changes in UV radiation due to different solar activity, incoming energetic particles and so on. Among the natural phenomena, energetic particles coming from the Sun or from outer space are able to ionize the air and trigger catalytic processes which can destroy O_3 involving NO_x and HO_x species, that are produced by this ionization (Crutzen, 1970; Krivolutsky et al., 2002; and references therein). When comparing the additional NO_x production through the energetic particles with e.g. the natural production by lightning at the southern hemispheric tropospherical polar region it can be seen that the percentage increase of more than 20 % caused by the energetic particles is more important than the additionally produced NO_x of less than 20 % through lightning (Legrand et al. 1988).
Ozone loss caused by solar proton events was initially observed by rocket measurements in 1969 (Weeks et al., 1972). Around the same time the first satellite observations of nitric oxide were made over polar regions (Rusch & Barth, 1975). Over the years several studies of solar proton event effects on the atmosphere have been published: The earlier work of Crutzen and Solomon (1980), McPeters et al. (1981), and Solomon et al. (1983) has been followed by several studies by Jackman and coworkers (Jackman and McPeters, 1985; Jackman and Meade, 1988; Jackman et al., 1990; Jackman et al., 1993; Jackman et al., 1995, 2000), McPeters and coworkers (McPeters

and Jackman, 1985; McPeters, 1986), and others (Reid et al., 1991; Callis et al., 1998).

The loss of ozone due to the Galactic cosmic rays (GCRs) has been modeled by Krivolutsky et al. (2002). Other papers analyzed the impact of the GCRs on the odd nitrogen, which in turn can destroy ozone (Nicolet, 1975; Vitt & Jackman, 1996).

The energetic electrons (EEPs) and their impact on ozone has been observed by Thorne (1977); (1980). Years later Baker et al. (1987) investigated the influence of the EEPs on the ozone loss in the Earth's atmosphere. During the next years, many studies on the effects of the energetic electrons on the atmospheric chemistry have been published (Callis et al., 1991; Gaines et al., 1995; Callis et al., 1998; Saetre et al., 2004; Rozanov et al., 2005; Randall et al., 2007; Funke et al., 2008; Clilverd et al., 2009).

Even though the influence of the GCRs, solar protons and energetic electrons on atmospheric chemistry and dynamics has been shown by several publications to be potentially significant, it is usually not properly taken into account for ozone trend analyses. For example the first Chemistry-Climate Model Validation Activity (CCMVal) for coupled chemistry climate models (CCMs) (Eyring et al., 2006) and in the most recent CCMVal report (see the homepage of SPARC, www.atmosp.physics.utoronto.ca/SPARC/CCMVAL_FINAL/index.php) ignores the effects of energetic particles on the atmosphere.

Seppälä et al. (2009) report by analyzing ERA-40 and ECMWF data that during winter months, polar surface air temperatures in years with high magnetic activity of the Sun, defined by the A_p index, differs from years with low A_p index. They show that the differences are statistically significant at the 2-sigma level which range up to about 4.5 Kelvin, depending on location.

Therefore, the implementation of GCRs, solar proton events (SPEs) and EEPs in 3-D CCMs is important to estimate the potential errors caused by neglecting the energetic particles.

To reach this goal the state-of-the-art CCM SOCOL v2.0 (Schraner et al., 2008) has been applied to simulate on the one hand the effects of the energetic particles separately and on the other hand when all effects are exerted simultaneously. Several new parameterizations have been included and used in modeling the impacts of the different particles on the Earth's atmosphere. Modeling the GCRs, SPEs and EEPs simultaneously in a 3-D CCM has not been performed before.

1.2 Objectives and outline

The previous section presented the topic of this thesis, the influence of the energetic particles on atmospheric chemistry and dynamics, in a broader context. The simulations with the GCRs, SPEs and EEPs will provide useful insights in chemical and dynamical processes and contribute to a better understanding on how the particles alter our atmosphere.

Thus, the main goals of this thesis are:

• **apply and/or develop parameterizations for GCRs, SPEs and EEPs**. This is to ensure an optimum fit of our model and the parameterizations for the energetic particles.

• **perform a number of simulations from ground to the mesopause with each class of the energetic particles and compare the results with satellite data and/or data from other models.** These simulations show where the GCRs, SPEs and EEPs have their most intense interaction with the Earth's atmosphere and show also the intensity of the downward propagation of NO_x and HO_x produced through the ionization.

• **perform and analyze a model run with all classes of energetic particles employed simultaneously**. The analysis of this run will provide a better understanding and clarify whether it is necessary to implement the GCRs, SPEs and EEPs in further model runs that deal with atmospheric chemistry and dynamics.

Introduction

The thesis is structured into 8 chapters with the following content:

Chapter 1 is giving a short introduction on the background of this thesis. It also highlights what the goals of this thesis are.

Chapter 2 will give an overview on the Galactic cosmic rays, the solar protons and the energetic electrons, where they are originating from, their energy spectra and how they interact with the Earth's atmosphere.

Chapter 3 will discuss the chemistry and the dynamics related to the energetic particles. In this chapter, the most important ozone destruction cycles are discussed and also the importance of the heterogenous chemistry will be highlighted.

Chapter 4 gives the model description and the experimental setup for the different model runs used within this thesis

Chapters 5 to 7 are devoted to results obtained during this study. Chapter 5 presents the results for the Galactic cosmic rays, whereas the next chapter analyzes the influence of a Carrington like solar proton event. Chapter 7 will present results where a run has been performed including all energetic particle classes simultaneously.

Chapter 8 summarizes the work presented and gives an outlook on subsequent projects.

Appendix A presents results from the HEPPA (High-Energy Particle Precipitation in the Atmosphere) community which focuses on observational (satellite) as well as modelling studies of atmospheric and ionospheric changes caused by energetic particle precipitation, e.g. solar proton events, relativistic electron precipitation, and auroral electron precipitation.

Chapter 2

Energetic particles

Several types of energetic particles penetrate into and interact with the Earth's atmosphere. They all have in common that they can ionize the constituents in the Earth's atmosphere.
On their way through the atmosphere, energetic particles collide with neutral atmospheric constituents and release part of their energy through the process of ionization. Thereby, additional NO_x and HO_x is produced which can lead to the destruction of ozone.
The present thesis studies the impact of three different particle types on the Earth's atmosphere: Galactic cosmic rays, solar protons and energetic electrons.

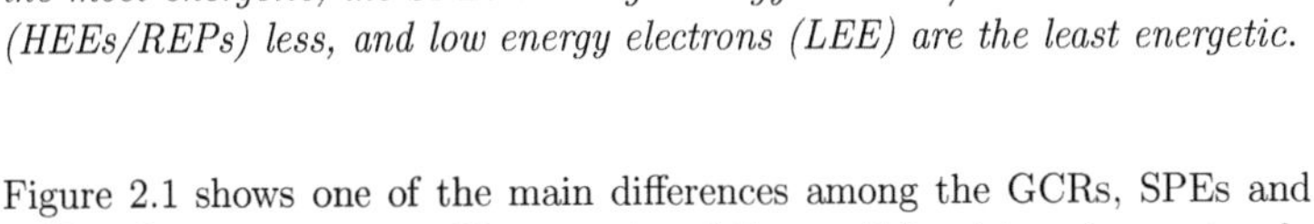

LEE	HEE/REP	SPEs	GCRs	Energy range
a few keV (Ho et al. 2003)	> 1 MeV/ > 500 keV (Pesnell, 2001)	10 Mev – 10 GeV (Makhmutov et al. 2005)	up to 10^{13} MeV (Bazilevskaya et al. 2008)	

Figure 2.1: *Energy range for the different energetic particles. The GCRs are the most energetic, the SPEs and high energy electrons/relativistic electrons (HEEs/REPs) less, and low energy electrons (LEE) are the least energetic.*

Figure 2.1 shows one of the main differences among the GCRs, SPEs and EEPs, the energy range. The energies of the particles determines not only the penetration depth (see Fig. 2.2), but also the geographical (latitude) region where the particles can enter the atmosphere. Figure 2.2 depicts also that electrons which have been strongly accelerated, so-called relativistic electrons (REPs), can have energies up to a few MeV so that they are able to penetrate down to the upper stratosphre.

Energetic particles

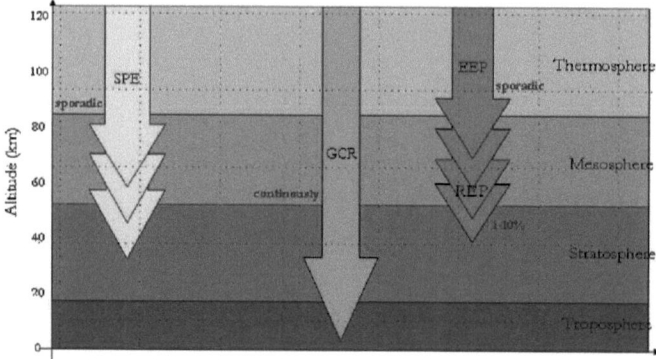

Figure 2.2: *Altitude range where the Galactic cosmic rays (GCRs), solar protons (SPEs) and energetic electrons (EEPs) interact with the Earth's atmosphere.*

Another difference among the particles is that they have different occurrence probabilities, i.e. the Galactic cosmic rays interact all the time with the Earth's atmosphere whereas the SPEs and EEPs show an irregular occurrence, depending on the solar activity. Due to their charge, GCRs, SPEs and EEPs are deflected by the geomagnetic field of the Earth. Almost all particles can penetrate into the polar region, whereas only the energetic particles with energies above 15 GeV are able to penetrate perpendicular to the magnetic field near the equator (Usoskin et al., 2004).

The following sections present the characteristics of the GCRs, SPEs and EEPs in more detail. They explain where the charged energetic particles originate from, where they are able to penetrate into the Earth's atmosphere and how and at which altitude range they ionize the air.

2.1 Galactic Cosmic Rays

2.1.1 Origin and characteristics

Galactic cosmic rays are high-energy particles (up to many GeV) that enter our solar system from far away in the Galaxy. GCRs consist mostly of protons and α - particles (Bazilevskaya et al., 2008). α - particles consist of two protons and two neutrons forming a particle identical to a Helium nucleus (He^{2+}).

Galactic cosmic rays are most probably accelerated in the blast waves of supernova remnants (http://helios.gsfc.nasa.gov/gcr.html). This does not mean that the supernova explosion itself gets the particles up to these speeds, but they could also be accelerated by shock waves when the expanding bubble of the hot super nova remnant gas encounters the interstellar medium.

The origin of the GCRs can only be determined by indirect means. One of the indirect observations we can make, the composition of the GCRs, can tell a lot about the sources, i.e. all elements heavier than He are reminder of Supernova explosions. All of the natural elements in the periodic table are present in GCRs, in roughly the same proportion as they occur in the solar system (Grieder, 2001). But detailed differences provide a fingerprint of the Galactic cosmic rays' source.

Figure 2.3 reveals that the intensity near the Earth shows the well-known inverse relationship to the 11-year Sunspot cycle (Neher, 1956; Usoskin et al., 2009).

This means that during high solar activity the intensity of the Galactic cosmic rays is about 30 % less than in times with low solar activity. This is caused by the interplanetary medium which has a larger shielding effect on the GCRs during high solar activity (Thorne, 1980; Svensmark & Friis-Christensen, 1997).

Energetic particles

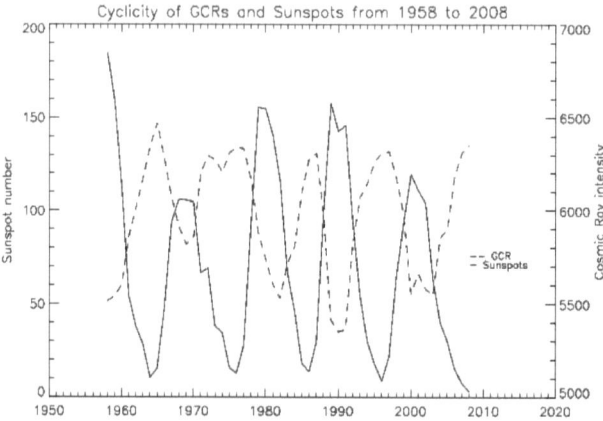

Figure 2.3: *Cyclicity of GCRs, taken from Kiel Neutron Monitor (http://cr0.izmiran.rssi.ru/kiel/main.htm) and Sunspots from 1958 to 2008.*

2.1.2 Interaction with Earth's atmosphere

As mentioned above, GCRs lead to additonal NO_x and HO_x production on their way through the atmosphere. Additionally, the GCRs can produce secondary particles which can be energetic enough to contribute themselves to further ionization of the neutral gases (Bazilevskaya et al., 2008). This leads to the development of an ionization cascade or primary cosmic ray particles (Bhabha & Heitler, 1937; Bridge et al., 1953). Cascades of particles with several hundred MeV of kinetic energy may reach the ground, and therefore may easily be detected by many types of particle detectors, e.g. neutron monitors.

Between a few km and 25-30 km, the cosmic ray induced ionization is the main source of the atmospheric ionization (Bazilevskaya & Svirzhevskaya, 1998) with the maximum rate due to the Bragg peak in the stopping power around 15 km, which can be recognized in Figure 2.4. About 90 % of the Galactic cosmic rays are protons, 9 % are α - particles (Grieder, 2001). The remaining 1 % consists of all other heavier elements and isotopes.

2.2 Solar Proton Events

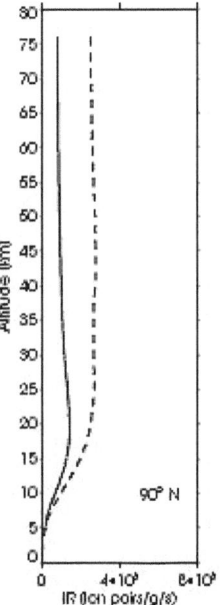

Figure 2.4: *Ionization rate (IR, number of ion pairs produced per air mass and time unit) for 90° geomagnetic latitude as computed by Usoskin, Kovaltsov & Mironova, (2010). Solid lines show IR during solar maximum and dashed lines during solar minimum.*

2.2 Solar Proton Events

2.2.1 Origin and characteristics

A solar proton event, which lasts for a few days, occurs when protons in the solar plasma are being emitted by the active Sun are accelerated to very high energies (up to 500 MeV) either close to the Sun during a solar flare or in interplanetary space by the shocks associated with coronal mass ejections. Solar flares mostly occur in active regions around sunspots. If a solar flare

Energetic particles

is exceptionally powerful, it can cause coronal mass ejections. At the peak of the solar cycle there are typically more sunspots on the Sun, and hence more solar flares. The frequency of occurrence of solar flares can vary from several per day when the sun is active, to less than one in one week during quiet phases (Gosling, 1993).

Solar protons normally have insufficient energy to penetrate through the Earth's magnetic field. However, during unusually strong solar flare events, protons can be produced with sufficient energies to penetrate deeper into the Earth's atmosphere (Shea et al., 2005).

When solar protons enter the domain of the Earth's magnetosphere, where the magnetic fields become stronger than the solar magnetic fields, they are then guided by the magnetic field of the Earth into the polar regions where the majority of the field lines enter deep into the atmosphere (Fig. 2.5).

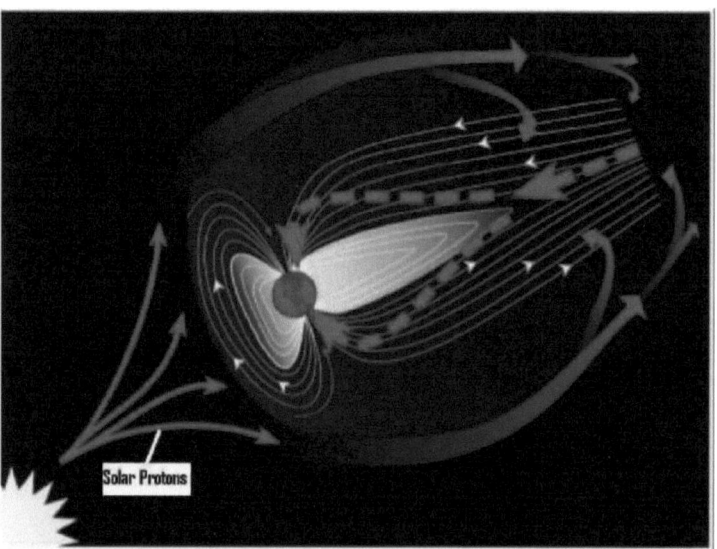

Figure 2.5: *Pathway of the Solar Protons to the Earth. (Taken from http://Sunclimate.gsfc.nasa.gov/projects/)*

2.2 Solar Proton Events

2.2.2 Interaction with Earth's atmosphere

In contrast to Galactic cosmic rays, which can penetrate the Earth's atmosphere at all latitudes, the solar protons enter the atmosphere between 60° and 90° geomagnetic latitude due to the reduced geomagnetic field shielding (Bazilevskaya et al., 2008; Damiani et al., 2009). Depending on the energy spectra of the solar protons, their largest ionization rate can be found in the mesosphere (around 50-80 km) as shown in Fig. 2.2.

The enhanced ionization due to incoming energetic protons at the polar cap ionosphere level can have the effect of completely blocking all radio waves passing through this heavily ionized region (Yoshida, 1965). Such events are known as Polar Cap Absorption events (PCAs). These events commence and last as long as the energy of incoming protons at approximately greater than 10 MeV exceeds roughly 10 pfu (particle flux units) at geosynchronous satellite altitudes (Rose & Ziauddin, 1962).

A common way to detect solar proton events is using satellites, i.e. the satellites carry detectors which are able to quantify the flux of the solar protons (European Space Agency, 2000). Before the satellites era, ionization chambers have been used for this purpose. A detailed description of how such ionization chambers work can be found in Rossi & Staub (1949).

2.3 Energetic Electron Precipitation

2.3.1 Origin and characteristics

The Van Allen radiation belt (named after James Van Allen) is a torus of energetic charged particles around Earth, which is fed mostly by solar particles and held in place by the Earth's magnetic field. This field is not uniformly distributed around the Earth. On the sunward side, it is compressed by the solar wind. The Van Allen belt is split into two distinct belts, with energetic electrons forming the outer belt and a combination of protons and electrons creating the inner belt (Fig. 2.6).

The large outer radiation belt extends from an altitude of about three to ten Earth radii (RE) above the Earth's surface, and its greatest intensity is usually around 4-5 RE. The outer belt consists mainly of high energy electrons (0.1 up to 10 MeV) trapped by the Earth's magnetosphere. They are injected from the geomagnetic tail following geomagnetic storms, and are subsequently energized through wave-particle interactions. Particles are trapped in the Earth's magnetic field because it is basically a magnetic mirror. Particles gyrate around field lines and also move along field lines. As particles encounter regions of stronger magnetic field where field lines converge, their longitudinal velocity is slowed and can be reflected. This causes the particle to bounce back and forth between the Earth's poles, where the magnetic field increases (Hudson et al., 2008).

The inner Van Allen Belt spans between an altitude of 700 - 10,000 km (0.1 to 1.5 Earth radii) above the Earth's surface containing mostly energetic protons with energies up to 100 MeV and electrons in the range of hundreds of keV (Tascione, 1994).

2.3 Energetic Electron Precipitation

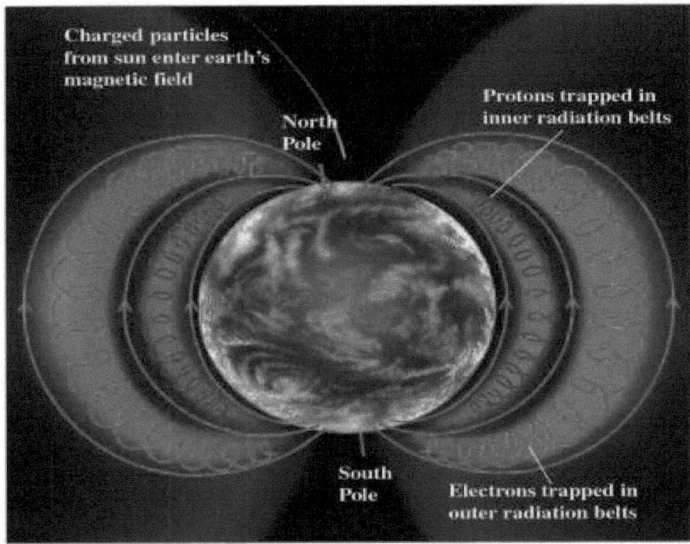

Figure 2.6: *Earth with the Van Allen Belts and the trapped electrons. Taken from http://www.physics.sjsu.edu/becker/physics51/mag*

Energetic electron fluxes can increase and decrease dramatically as a consequence of geomagnetic storms, which are themselves triggered by magnetic field and plasma disturbances produced by the Sun. The increases are due to storm-related injections and acceleration of particles from the tail of the magnetosphere. Measurements show that energetic electron events are more frequent and intense during the declining phase of the solar activity, when coronal holes migrate towards the solar equator and the solar wind is more directed toward the Earth (Callis et al. 1998; Bazilevskaya et al. 2002).

The energy spectra of the electrons can vary, depending how strong they have been accelerated during the geomagnetic storm. Low energetic electrons are in the range of 1 - 100 keV. High energetic electrons reach values up to 2 MeV (e.g., Gaines et al.,1995). Electrons with energies higher than 0.8 MeV are so called relativistic electrons. Only 1 to 10 % of all energetic electrons penetrating the Earth's atmosphere are relativistic electrons (Callis et al.

1991).

2.3.2 Interaction with Earth's atmosphere

The accelerated electrons can, after precipitation, penetrate into the atmosphere over the auroral and sub-auroral regions (see Fig. 2.7) and ionize neutral components of the atmosphere, similar to GCRs and SPEs, providing a source of reactive nitrogen and hydrogen which can destroy ozone (Hudson et al., 2008; Lopez & Baker, 1994; Rozanov et al., 2005).

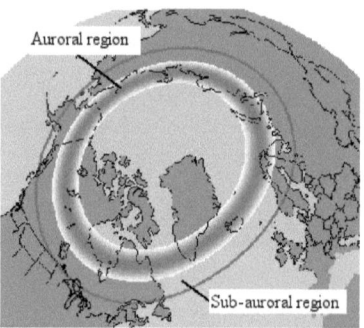

Figure 2.7: *Earth showing the auroral and sub-auroral regions. Taken from http://odin.gi.alaska.edu/FAQ/*

Depending on the energy level of the electrons, the penetration depth in the Earth's atmosphere can vary from more than 100 km down to approximately 65 km. The majority of the energy is deposited in the lower to middle thermosphere which is around 80-100 km in altitude (Callis et al., 1998).

2.3 Energetic Electron Precipitation

ε_0 keV	$\lambda_m = 49°$			$\lambda_m = 67°$			$\lambda_m = 75°$		
	10% (km)	max (km)	10% (km)	10% (km)	max (km)	10% (km)	10% (km)	max (km)	10% (km)
0·10	236	288	550	231	282	540	231	282	540
0·20	206	252	495	202	246	485	202	246	480
0·30	188	225	450	184	222	445	183	222	440
0·50	166	198	390	164	192	385	163	192	380
0·70	154	176	350	151	174	340	152	172	340
1·00	142	156	299	141	156	291	140	156	290
1·50	133	144	248	131	142	242	131	142	240
2·00	128	138	217	125	134	212	126	134	212
3·00	120	128	184	119	126	180	119	126	180
5·00	111	118	156	111	118	154	111	118	154
10·00	103	108	135	102	108	135	102	108	134
17·00	97	102	124	96	100	124	96	100	124
30·00	91	96	116	91	94	116	90	94	115
50·00	85	90	108	85	90	108	85	90	108
100·00	79	84	100	79	84	100	79	84	100
200·00	71	78	94	71	76	93	71	76	94
300·00	67	74	90	67	72	90	67	72	90
500·00	61	68	86	61	66	86	61	66	86

Table 2.1: *Altitude at which electrons of a given energy produce the maximum ionization rate for several geomagnetic latitudes (λ_m). The heights at which the ionization rates have decreased to 10 % below and above the maximum are also given in this table. (Taken from M. H. Rees, 1964)*

Table 2.1 shows the altitude at which electrons of a given energy produce the maximum ionization rate for several geomagnetic latitudes.

One way to detect energetic electrons is via satellite measurements, or by ground-based VLF (Very Low Frequency) transmitters. A detailed description of ground-based VLF transmitters and how they work can be found in Barr et al. (2000).

Energetic particles

Chapter 3

Chemistry and dynamics

This chapter discusses the chemical and physical processes, mainly related to ozone from ground to the upper mesosphere. In particular, chemical and physical processes that are related to this thesis will be treated, i.e. the ionization of the neutral atmosphere through the energetic particles. For the realisation of this chapter, several textbooks, e.g. Brasseur and Solomon (2005) have been used.

3.1 The role of Ozone and its chemistry

Ozone plays a key role in the chemical and radiative budgets of the middle atmosphere. By absorbing solar energy at wavelengths shorter than approximately 320 nm, it protects the biosphere from harmful solar UV-radiation, and from DNA damage in living cells from animals and human beings (Norval et al., 2007).
The first theory for the formation and destruction of stratospheric ozone involving only oxygen reactions was presented by the British geophysicist S. Chapman. The following reactions present the major photochemical processes affecting the formation and destruction of ozone in a pure oxygen atmosphere. In the middle atmosphere, chemistry related to ozone is frequently described in terms of the odd oxygen familiy, O_x. This family consist of the ground state oxygen atom, $O(^3P)$, the first exited state $O(^1D)$ and the ozone molecule, O_3. Figure 3.1 presents the main chemical reactions for the odd

oxygen family.

$$O_2 + h\nu \rightarrow O(^3P) + O(^3P) \qquad \lambda < 242.4\text{nm} \qquad (3.1)$$
$$O(^3P) + O(^3P) + M \rightarrow O_2 + M \qquad (3.2)$$
$$O(^3P) + O_2 + M \rightarrow O_3 + M \qquad (3.3)$$
$$O(^3P) + O_3 \rightarrow 2O_2 \qquad (3.4)$$
$$O_3 + h\nu \rightarrow O_2 + O(^3P) \qquad \lambda \geqslant 320\text{nm} \qquad (3.5)$$
$$O_3 + h\nu \rightarrow O_2 + O(^1D) \qquad \lambda \leqslant 320\text{nm} \qquad (3.6)$$

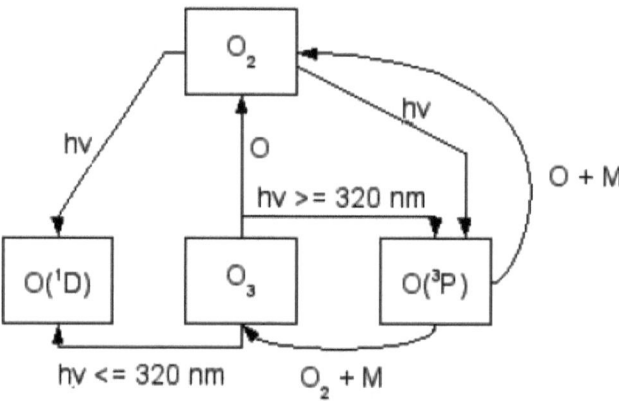

Figure 3.1: *Main chemical reactions for the O_x family*

Equation (3.1) represents the photodissociation of molecular oxygen by ultraviolet radiation at wavelengths less than 242.4 nm which produces atomic oxygen. This atomic oxygen can react in a three body process (3.2), whereas M represents an N_2 or an O_2 which takes the energy that has been released by the binding energy. The reaction (3.2) plays a significant role only in the thermosphere, where the concentration of atomic oxygen is large. (3.3) is the only known reaction in the atmosphere which leads to ozone production. The reaction seen in (3.4) is strongly temperature dependent and provides a loss for the odd oxygen familiy, producing O_2. The photodissociation of

3.1 The role of Ozone and its chemistry

ozone leads to formation of oxygen atoms in either ground state (3.5) or first exited state (3.6).

3.1.1 The distribution of ozone in the atmosphere

Most of the ozone molecules are found in a vertical column ranging from 10 km to 35 km. It is well known that ozone has seasonal and latitudinal variations with a maximum in its column in the region of the least production. The global distribution of total ozone (i.e. column ozone) as a function of latitude and time, taken from TOMS, is shown in Figure 3.2 given in Dobson Units (DU), which correspond to the thickness that the ozone column would have if the pure gas was compressed to standard pressure and temperature (1 DU = 2.69 * 10^{16} molecules cm^{-2}, i.e. 100 DU correspond to a 1mm thick layer of pure ozone). This Figure shows that the ozone column in the tropics is smaller than in the higher latitudes.

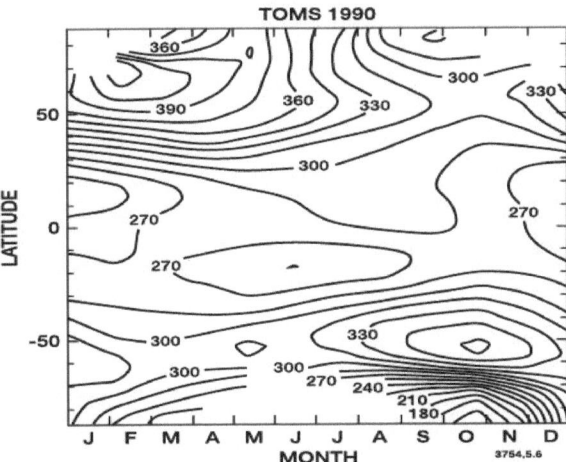

Figure 3.2: *Global map of the variations of total ozone, given in DU, measured by the TOMS instrument in 1990 (taken from Brasseur and Solomon, 2005.)*

The very low ozone abundance visible over the Antarctic from September to November is referred to as the ozone hole. During the 1990s and the first decade of the new millenium, low ozone levels have also been recorded during the Arctic spring. These anomalous ozone levels are associated with cold temperatures in the polar lower stratosphere and the activation of chlorine on surfaces of polar stratospheric cloud particles (see chapter about heterogenous chemistry). In the stratosphere, the production of ozone occures mainly where the solar radiation is the most intense, i.e. in the tropics at about 30 to 40 km. The Brewer-Dobson-Circulation causes ozone that is produced in the tropics to be transported poleward and downward, which is pronounced in the winter hemisphere (Fig. 3.3).

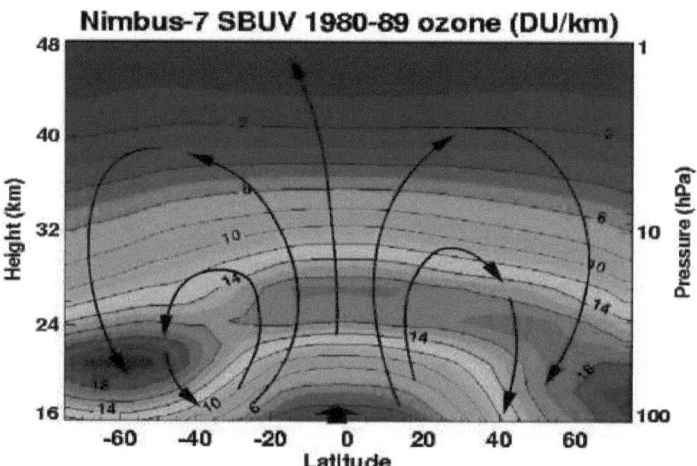

Figure 3.3: *Ozone number density in DU per kilometer averaged over a 10 year period. The data is based on the measurements of the Nimbus-7 SBUV instrument from 1980-1989. The black arrows in the figure represent the annual average of the Brewer-Dobson-Circulation in the stratosphere. Image taken from http://www.ccpo.odu.edu/~lizsmith/SEES/ozone/oz_class.htm*

3.2 NOx chemistry

The presence of NO_x results from the oxidation of nitrous oxide (N_2O) and, through ionization or dissociative ionization of molecular nitrogen (N_2) by energetic particles. The following paths show how NO_x can be produced via the energetic particles (Aikin, 1994; Egorova et al., 2010). The reactions lead to the formation of both ground state nitrogen $N(^4S)$ and excited state nitrogen $N(^2D)$.

$$N_2 + e \rightarrow N^+ + N(^4S) + 2e \qquad (3.7)$$
$$N_2 + e \rightarrow N(^4S) + N(^2D) + e \qquad (3.8)$$
$$N^+ + N_2 \rightarrow N_2^+ + N(^4S) \qquad (3.9)$$
$$N(^2D) + O_2 \rightarrow NO + O \qquad (3.10)$$
$$N(^4S) + O_2 \rightarrow NO + O \qquad (3.11)$$

In contrast to these ionization-related processes, the main pathway to odd nitrogen production in the stratosphere is through the reaction with excited oxygen (3.12 - 3.13). Nitric oxide (NO) and nitrogen dioxide (NO_2) are in photochemical balance during daytime due to fast conversion mechanisms, i.e. NO is converted to NO_2 by reaction with ozone, and NO_2 is transformed back to NO either by photolysis or by reaction with atomic oxygen (3.17 - 3.19). Photodissociation of nitric oxide initiates the primary loss process for NO_x in the middle atmosphere (3.15), because it leads to the formation of ground-state nitrogen atoms $N(^4S)$, which can react with NO in a so-called cannibalistic reaction (3.16).

$$N_2O + O(^1D) \rightarrow 2NO \qquad (3.12)$$
$$N_2O + O(^1D) \rightarrow N_2 + O_2 \qquad (3.13)$$
$$N_2O + h\nu \rightarrow N_2 + O(^1D) \qquad \lambda \leqslant 200\text{nm} \qquad (3.14)$$
$$NO + h\nu \rightarrow N(^4S) + O \qquad (3.15)$$
$$N(^4S) + NO \rightarrow N_2 + O \qquad (3.16)$$
$$NO + O_3 \rightarrow NO_2 + O_2 \qquad (3.17)$$
$$NO_2 + h\nu \rightarrow NO + O(^3P) \qquad \lambda < 405\text{nm} \qquad (3.18)$$
$$NO_2 + O \rightarrow NO + O_2 \qquad (3.19)$$

Chemistry and dynamics

The ozone destruction caused by the NO_x production cycles (via energetic particles or N_2O) is shown in the following pathway.

NO_x-Cycle 1, (Crutzen 1970)

$$NO + O_3 \rightarrow NO_2 + O_2 \quad (3.20)$$
$$\underline{NO_2 + O \rightarrow NO + O_2} \quad (3.21)$$
$$Net : O_3 + O \rightarrow 2O_2 \quad (3.22)$$

This cycle is most efficient near 35 to 45 km. Above this range, this cycle slows because the concentration of ozone is decreasing. The same is true below 35 km, where the concentration of O decreases and therefore reaction 3.21 is getting slower.

NO_x-cycle 2

$$NO + O_3 \rightarrow NO_2 + O_2 \quad (3.23)$$
$$NO_2 + O_3 \rightarrow NO_3 + O_2 \quad (3.24)$$
$$\underline{NO_3 + h\nu \rightarrow NO + O_2} \quad (3.25)$$
$$Net : 2O_3 \rightarrow 3O_2 \quad (3.26)$$

This cycle is more important in the lower stratosphere because O radicals are not needed. NO_3 can, instead of making photolysis, react with NO_2 which is shown in the following cycle.

There are other cycles involving the N_2O_5 molecule, which are neglected here (see Brasseur & Solomon, 2005).

Note that this cycle results in ozone destruction only when the photolysis of NO_3 leads to $NO + O_2$ rather than $NO_2 + O$. In the latter case, the effect is null. Besides the ozone destruction through NO_x species in the middle atmosphere, a reaction chain producing ozone, depending on the concentration of NO, exists in the troposphere (3.27 - 3.31). Because GCRs are able to penetrate down to the surface they are able to modulate the concentration of NO in the troposphere and therefore influence the production of ozone.

$$CO + OH \rightarrow CO_2 + H \quad (3.27)$$
$$H + O_2 \rightarrow HO_2 + M \quad (3.28)$$
$$HO_2 + NO \rightarrow OH + NO_2 \quad (3.29)$$
$$NO_2 + h\nu \rightarrow NO + O \quad (3.30)$$
$$\underline{O_2 + O \rightarrow O_3 + M} \quad (3.31)$$
$$Net : CO + 2O_2 \rightarrow CO_2 + O_3 \quad (3.32)$$

3.3 HOx chemistry

The production of HO_x is maintained by the dissociation of H_2O, CH_4 and H_2 which varies strongly with altitude. The main source of HO_x in the mesosphere and above is the (UV) photolysis of water vapour. In the stratosphere and lower mesosphere HO_x is formed in the reaction of water vapour with $O(^1D)$. Methan and H_2 react also with $O(^1D)$ to produce HO_x species.
HO_x is also produced through precipitation of energetic particles. The initial production of ion pairs is followed by formation and reactions of water cluster ions which result in the production of HO_x. The main processes responsible for the HO_x formation were considered by Solomon et al. (1981) and Aikin (1994).
Equations 3.33 and 3.34 show the production path of O_2^+ and O_4^+.

$$O_2 + e \rightarrow O_2^+ + 2e \qquad (3.33)$$
$$O_2^+ + O_2 + M \rightarrow O_4^+ + M \qquad (3.34)$$

The following cycles (3.35 - 3.38) show an example how O_2^+ and O_4^+ further react to produce HO_x.

$$O_2^+ + H_2O \rightarrow O_2^+H_2O \qquad (3.35)$$
$$O_2^+H_2O + H_2O \rightarrow H^+H_2O + OH + O_2 \qquad (3.36)$$
$$H^+H_2O + e \rightarrow H_2O + H \qquad (3.37)$$
$$O_4^+ + H_2O \rightarrow O_2^+H_2O + O_2 \qquad (3.38)$$

The H atom formed in (3.37) can recombine with O_2 according to (3.28) and generate additional HO_x Because of the formation of negative ions, dissociative recombination is not important at altitudes below 70 km. Instead ion-ion recombination becomes operative which is shown in the following paths (Aikin, 1994).

$$NO_3^- + H^+H_2O \rightarrow HNO_3 + H_2O \qquad (3.39)$$
$$HNO_3 + h\nu \rightarrow OH + NO_2 \qquad (3.40)$$

Chemistry and dynamics

The pathway to HO_x production through the reactions of H_2O, CH_4 and H_2 is presented in 3.41 - 3.44.

$$H_2O + h\nu \rightarrow H + OH \qquad \lambda < 200\text{nm} \qquad (3.41)$$
$$H_2O + O(^1D) \rightarrow 2OH \qquad (3.42)$$
$$CH_4 + O(^1D) \rightarrow CH_3 + OH \qquad (3.43)$$
$$H_2 + O(^1D) \rightarrow OH + H \qquad (3.44)$$

The HO_x species are very reactive and thus have a very short chemical lifetime in the atmosphere. Below 80 km the lifetime of the HO_x family is of the order of minutes to hours. Therefore, the HO_x distribution is almost independent of transport processes. Above 80 km the chemical lifetime of HO_x increases but because of the low abundance of water vapour the concentration of HO_x at these altitudes is negligible.

The interaction of HO_x with ozone is altitude dependent, i.e. several cycles exist how ozone can be destroyed by hydrogen radicals. The first cycle must be considered in the middle and lower stratosphere and in the troposphere.

HO_x-cycle 1

$$OH + O_3 \rightarrow HO_2 + O_2 \qquad (3.45)$$
$$\underline{HO_2 + O_3 \rightarrow 2O_2 + OH} \qquad (3.46)$$
$$\text{Net}: 2O_3 \rightarrow 3O_2 \qquad (3.47)$$

At altitudes between 30 km and 40 km, where more atomic oxygen is abundant, the reactions (3.48 - 3.50) are important, whereas above 40 km, the reaction of OH with O is faster than with ozone (3.51).

HO_x-cycle 2

$$OH + O_3 \rightarrow HO_2 + O_2 \qquad (3.48)$$
$$\underline{HO_2 + O \rightarrow O_2 + OH} \qquad (3.49)$$
$$\text{Net}: O_3 + O \rightarrow 2O_2 \qquad (3.50)$$

$$OH + O \rightarrow O_2 + H \qquad (3.51)$$

The H radical from 3.51 mostly reacts with O_2 which produces HO_2 (3.52). Instead of reacting with O_2, the H radical can directly react to OH with reaction with O_3 (3.53).

$$H + O_2 + M \rightarrow HO_2 + M \qquad (3.52)$$
$$H + O_3 \rightarrow OH + O_2 \qquad (3.53)$$

3.4 Halogen chemistry

3.4.1 Chlorine

The natural production of chlorine atoms in the stratosphere is provided by the destruction of methyl chloride, either by photolysis or by reaction with the hydroxil radical (3.54 - 3.55).

$$CH_3Cl + h\nu \rightarrow CH_3 + Cl \qquad (3.54)$$
$$CH_3Cl + OH \rightarrow CH_2Cl + H_2O \qquad (3.55)$$

The major source of anthropogenic chlorine results from the photolysis of chlorofluorocarbons (CFCs) and other chlorocarbons in the stratosphere (3.56).

$$CFCl_3 + h\nu \rightarrow CFCl_2 + Cl \qquad \lambda < 226\text{nm} \qquad (3.56)$$

The importance of the CFCs regarding the ozone destruction has been noted by Molina & Rowland, 1974, which is represented in the first cycle.

ClOx-cycle 1

$$Cl + O_3 \rightarrow ClO + O_2 \qquad (3.57)$$
$$ClO + O \rightarrow Cl + O_2 \qquad (3.58)$$
$$O_3 + h\nu \rightarrow O_2 + O \qquad (3.59)$$
$$\text{Net} : 2O_3 + h\nu \rightarrow 3O_2 \qquad (3.60)$$

When high ClO concentrations are present in the stratosphere, i.e. during polar night when air masses are processed by polar stratospheric clouds (PSCs), the following cycle, where ClO reacts with itself, can start.

ClOx-cycle 2, (Molina & Molina, 1987)

$$2(Cl + O_3 \rightarrow ClO + O_2) \qquad (3.61)$$
$$ClO + ClO + M \rightarrow Cl_2O_2 + M \qquad (3.62)$$
$$Cl_2O_2 + h\nu \rightarrow 2Cl + O_2 \qquad (3.63)$$
$$\text{Net} : 2O_3 + h\nu \rightarrow 3O_2 \qquad (3.64)$$

3.4.2 Bromine

The natural production of bromine atoms is made by methyl bromide (CH_3Br) which is released by the ocean and is also released by biomass burning. Cycle 1 is mostly present in the stratosphere below 40 km.

Bromine-cycle 1

$$Br + O_3 \rightarrow BrO + O_2 \quad (3.65)$$
$$\underline{BrO + O \rightarrow Br + O_2} \quad (3.66)$$
$$Net : O_3 + O \rightarrow 2O_2 \quad (3.67)$$

During polar night, bromine can also react with itself (3.68 - 3.70).

Bromine-cycle 2

$$2(Br + O_3 \rightarrow BrO + O_2) \quad (3.68)$$
$$\underline{BrO + BrO + M \rightarrow 2Br + O_2} \quad (3.69)$$
$$Net : 2O_3 \rightarrow 3O_2 \quad (3.70)$$

Another important reaction that can destroy ozone has been noted by McElroy et al., (1992), which is illustrated in cycle 3.

Bromine-cycle 3

$$Cl + O_3 \rightarrow ClO + O_2 \quad (3.71)$$
$$Br + O_3 \rightarrow BrO + O_2 \quad (3.72)$$
$$ClO + BrO + M \rightarrow BrCl + O_2 + M \quad (3.73)$$
$$\underline{BrCl + h\nu \rightarrow Br + Cl \quad \lambda < 540nm} \quad (3.74)$$
$$Net : 2O_3 + h\nu \rightarrow 3O_2 \quad (3.75)$$

3.5 Heterogeneous chemistry

The cold temperatures that occur in polar winter can lead to formation of clouds within the stratosphere. The term polar stratospheric clouds was proposed by McCormick et al., (1982), who first presented satellite observations of high-altitude clouds in the Antarctic and Arctic stratospheres, but the clouds were considered little more than a scientific curiosity, until the ozone hole was discovered.

Solomon et al., (1986) suggested that HCl and ClONO$_2$ might react on the surfaces of PSCs which accelerates ozone loss. The Cl$_2$ formed would photolyze in Sunlit air and form ClO (3.61 - 3.64). Solomon et al., mentioned that this and related heterogeneous reactions would suppress the concentrations of NO$_2$ by forming HNO$_3$, so that ClO could not reform the ClONO$_2$ reservoir.

$$HCL + ClONO_2 \rightarrow HNO_3 + Cl_2 \tag{3.76}$$

Observations of PSCs, low NO$_2$ amounts in polar regions, enhanced polar HNO$_3$ and the vertical profile of the ozone depletion based upon the Japanese measurements (Chubachi, 1984) were cited in support of heterogeneous chemistry as the primary process initiating Antarctic ozone depletion. Such a mechanism would be most effective in the Antarctic due to colder temperatures and greater PSC frequencies there than in the corresponding seasons in the Arctic (McCormick et al., 1982).

These days, we know that heterogeneous reactions are one of the fastest reactions in the stratosphere. Equations (3.77 - 3.83) show the most important heterogeneous reactions.

$$ClONO_2 + HCl \rightarrow HNO_3 + Cl_2 \tag{3.77}$$
$$ClONO_2 + H_2O \rightarrow HNO_3 + HOCl \tag{3.78}$$
$$HOCl + HCl \rightarrow H_2O + Cl_2 \tag{3.79}$$
$$BrONO_2 + HCl \rightarrow HNO_3 + BrCl \tag{3.80}$$
$$BrONO_2 + H_2O \rightarrow HNO_3 + HOBr \tag{3.81}$$
$$HOBr + HCl \rightarrow BrCl + H_2O \tag{3.82}$$
$$N_2O_5 + H_2O \rightarrow 2HNO_3 \tag{3.83}$$

3.6 Dynamics and transport

This section gives a general description of the dynamical and the transport processes in the atmosphere, especially at the polar regions.

3.6.1 The Polar Vortex

During the hemispheric winter, the very cold air over the poles is surrounded by warmer air at lower latitudes. This creates a low pressure region with strong winds blowing around the region at the boundary between warm and cold air. The rotating air, a strong polar vortex, isolates the stratosphere over the poles from rest of the stratosphere (Fig. 3.4). The westerly circulation of

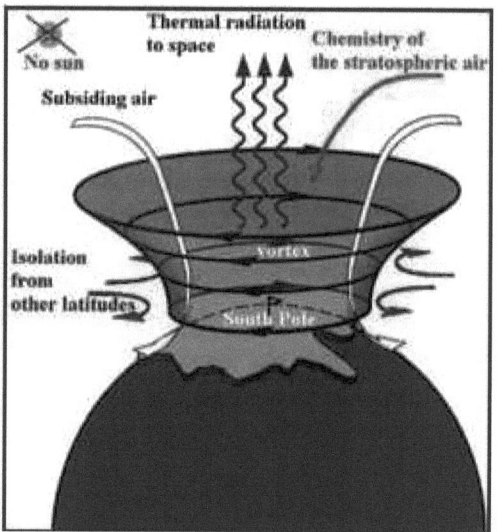

Figure 3.4: *Polar vortex and its consequences. Taken from Climate & Society Lectures at Columbia University Department of Earth and Environmental Sciences: Climate and Society*

the polar vortex is strongest in the upper stratosphere and strengthens over the course of the winter.

The polar night jet in the SH is important because it acts as a transport barrier between the southern polar region and the southern midlatitudes. It effectively blocks any mixing between air inside and outside the vortex during the winter. This means, that ozone-rich air from the midlatitudes can not be transported into the SH polar region. The isolation of polar air allows

3.6 Dynamics and transport

the ozone loss processes to proceed without any replenishment by intrusions of ozone-rich air from midlatitudes.

This isolation of the polar vortex is a key ingredient to polar ozone loss, since the vortex region can evolve without being disturbed by the more conventional chemistry of the midlatitudes.

The polar night jet over the Arctic is not as effective in keeping out intrusions of warmer, ozone-rich midlatitude air. This is because the north polar vortex is generally more disturbed by atmospheric waves forced beneath by flow over a more variable surface topography and hence more north-south mixing of air in the NH than in the SH. The winds in the polar vortex can reach maximum velocities of about 70 m/s near 40 km altitude. The edge of the vortex is usually near 60° N/S and it extends from approximately 16 km to the mesosphere.

3.6.2 The Brewer Dobson Circulation

The Brewer Dobson circulation, also known as the stratospheric meridional residual circulation, is driven by extratropical wave forcing in the middle atmosphere (Holton 1995). The circulation model suggested by Brewer (1949) and Dobson (1956) consists of three parts. The first part is rising tropical motion from the troposphere into the stratosphere, The second part is poleward transport in the stratosphere and the third part is descending motion in both the stratospheric middle and polar latitudes.

In the mesosphere, the meridional circulation is formed by a single cell in which rising motion takes place in the summer pole starting from the stratosphere, pole-to-pole transport in mesosphere-lower-thermosphere, and downward motion in the winter pole mesosphere, down to the stratosphere.

The Brewer Dobson circulation transports the ozone that is mainly produced in the tropics poleward and downward which leads to a global distribution of ozone which can be seen in Figure 3.3

3.6.3 Waves

Rossby waves are planetary waves generated in the troposphere by ocean-land temperature contrasts and topographic forcing (winds flowing over mountains), and affected by the Coriolis effect due to the earth's rotation. These waves propagate westward relative to the mean flow. Quasi-stationary waves have the largest amplitudes because they are produced by stationary sources including large-scale orography and land-sea contrasts. They propagate upwards and equatorwards in the middle atmosphere. Rossby waves propagate into the stratosphere primarly during winter. This propagation during winter time can get disturbed by the GCRs which can strengthen the zonal wind therefore alter the propagation of the planetary waves that can, at the end, influence the surface air temperature (Shindell et al., 2001).

Free Rosby waves with smaller amplitudes do propagate in the summer and can cause substantial mixing in the lower stratosphere at high latitudes (Wagner & Bowman, 2000). This could explain the variability of ozone observed at mid- to high latitudes during the summer in an altitude range of 20 to 30 km (Hoppel et al., 1999). Breaking of Rossby waves causes rapid irreversible mixing of the air parcels. Sudden stratospheric warming (SSW) can occur by dissipation of Rossby waves, producing the warming by decelerating the mean flow. These SSWs normally happen in the northern hemisphere (Holton, 2004).

Gravity waves are oscillations with relatively short horizontal wavelenghts (10 to 1000 km) that arises when air parcels are being displaced vertically. These waves are normally produced by air flow over mountains. The propagation of gravity waves through the atmosphere depends on the wind distribution and thermal structure. These waves provide the major source of dynamical variability in the mesosphere. Mesospheric gravity waves are predominantly westward propagating in winter and eastward in summer. They break at mesospheric altitudes, where the wave amplitude has grown so large that the vertical temperature perturbation results in the air parcels becoming convectively unstable i.e. they break. Gravity wave breaking controls the circulation of the mesosphere and the transport of trace gases between the stratosphere and mesosphere.

Chapter 4

Model description, parameterizations and experimental setup

4.1 Model description

The Chemistry Climate Model (CCM) SOCOL v2.0 (modeling tool for **SO**lar **C**limate **O**zone **L**inks studies) is a combination of the general circulation model (GCM) MA-ECHAM4 ant the chemistry transport model (CTM) MEZON. MA-ECHAM4 (Manzini et al., 1997) is a spectral model with T30 horizontal truncation resulting in a grid spacing of about 3.75°. In the vertical direction the model has 39 levels in a hybrid sigma-pressure coordinate system spanning the model atmosphere from the surface up to 0.01 hPa. A semi-implicit time stepping scheme is used with a time step of 15 min in the dynamical core. Physical processes parameterizations and full radiative transfer calculations are performed every 2 hours, but heating and cooling rates are updated every 15 minutes.

The chemical-transport part MEZON (Rozanov et al., 1999; Egorova et al., 2003) has the same vertical and horizontal resolution as the underlying GCM. It treats 41 chemical species of the oxygen, hydrogen, nitrogen, carbon, chlorine and bromine groups, which are determined by 140 gas-phase reactions, 46 photolysis reactions and 16 heterogeneous reactions in/on aqueous sulfuric acid aerosols, water ice and nitric acid trihydrate (NAT). The background tropospheric chemistry is included, however the model does not consider any non-methane hydrocarbons (NMHC) and volatile organic components (VOCs). The chemistry is driven by the prescribed mixing ratios of the source gases in the planetary boundary layer and prescribed sources of CO and NO_x. The heterogeneous chemistry scheme includes HNO_3 uptake by aqueous sulfuric acid aerosols (Carslaw et al., 1995) and a parameterization of the liquid-phase reactive uptake coefficients following Hanson & Ravishankara (1995) and Hanson et al. (1996). The PSC scheme for water ice uses prescribed particle number densities and assumes the cloud particles to be in thermodynamic equlibrium with their gaseous environment. NAT is formed if the partial pressure of HNO_3 exceeds its saturation pressure. A prescribed mean particle radius of $5\mu m$ is used for NAT, and the particle number densities are limited by an upper boundary of $5 * 10^{-4}$ cm^{-3} to take into account the fact that observed NAT clouds are often strongly supersaturated.

The transport scheme is a combination of the Prather scheme (Prather, 1986) applied for the vertical transport, and a semi-Lagrangian (SL) scheme, which is used for horizontal advection on a sphere (Williamson & Rasch, 1989). Since the SL scheme is not mass conserving, a mass fixer has to be applied after each advection step. Special modifications to the operational SL mass fixing routine are desribed in Schraner et al. (2008).

The GCM and CTM components of SOCOL are coupled via the three-dimensional fields of wind, temperature, ozone, water vapour, methane, nitrous oxide and chlorofluorocarbons (CFCs). The GCM provides the horizontal and vertical winds, temperature and tropospheric humidity for the CTM, which returns 3-D fields of the gas species mixing ratios to the GCM in order to calculate radiation fluxes and heating rates. The water cycle in the troposphere is treated in the GCM part of the model, while the water vapour chemistry, PSC formation, condensation and transport in the stratosphere and mesosphere are treated in the chemistry-transport part of the model. Unique water vapour field is transferred from GCM to CTM and back at every step. The comprehensive description and evaluation of the CCM SOCOL v2.0 is presented by Schraner et al. (2008).

4.2 Parameterizations

The goal of these parameterizations is to study how energetic particles interact with the Earth's atmosphere, i.e. to see how atmospheric chemistry and the dynamics change when interacting with the GCRs, SPEs and EEPs.

Since SOCOL does not explicitly consider ion chemistry, the effect of the energetic particles cannot be directly simulated. Therefore it is necessary to

4.2 Parameterizations

find a relationship between the amount of energetic particles and the NO_x and HO_x production rates.

Following Porter et al. (1976) 1.25 NO_x molecules are produced per ion pair, with 45 % of this NO_x yielding ground state atomic nitrogen and 55 % resulting in $N(^2D)$ with instantaneous conversion to NO.

The production of HO_x has been studied and tabulated by Solomon & Crutzen (1981) with a 1-D time-dependent model of neutral and ion chemistry. They parameterized the number of odd hydrogen particles produced per ion pair as a function of altitude and ionization rate for daytime, polar summer conditions of temperature, air density and solar zenith angle.

Therefore, we use the ionization rates (IR) in the CCM SOCOL to take into account the GCR, SPE and EEP induced production of NO_x and HO_x.

When comparing the changes for NO_x, HO_x and ozone obtained with the parameterization for the different energetic particles with Egorova et al. (2010) which used a 3-D CCM with complete ion chemistry, it turns out, that our results are in relativ good agreement compared to a model that uses full ion chemistry.

4.2.1 GCR runs

With these parameterizations, we wanted to compare and find out if and how the results change when the ionization stops at 18 km and when it goes down to the surface. It is important to check if the odd nitrogen that is produced in the stratosphere through the parameterization by Heaps (1978) shows the same impact in the troposphere when transported down as modeled by Usoskin et al. (2010) that shows NO_x production down to the surface.

Heaps parameterization

The intensity of the Galactic cosmic rays depends on altitude, latitude and solar activity. The altitude plays a role because depending on the energetic level of the proton, not all particles are able to penetrate down to the same height, i.e. the higher the particles' energy, the deeper they penetrate into the atmosphere. The latitude is important because of the magnetosphere of the Earth. Figure 4.1 shows that the closer to the poles, the weaker is the shielding effect of the magnetosphere caused by the fact that the fieldlines are entering the Earth.

Model description, parameterizations and experimental setup

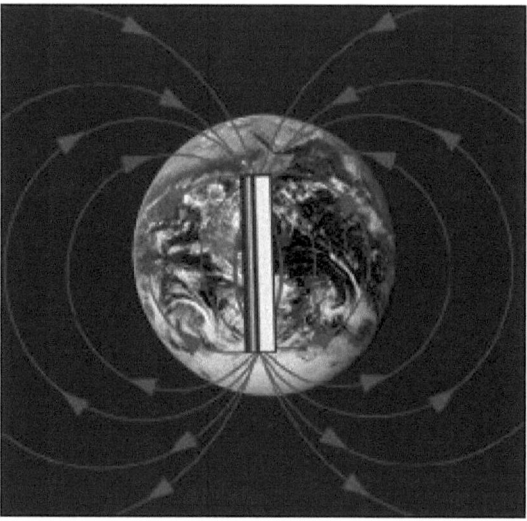

Figure 4.1: *The Earth's magnetic field is similar to that of a magnetic dipol. Clearly visible the entering of the lines close to the poles. Picture taken from ds9.ssl.berkeley.edu/themis/mission_magnetosphere.html*

The GCRs show an inverse relationship with the solar activity, i.e. during times when the Sun is active, the intensity of the GCRs is weakened and vice versa. The reason for this phenomena is because the solar activity influences the Interplanetary Magnetfield (IMF) which acts as a shielding against the energetic particles. These important dependencies has been taken into account for the parameterization proposed by Heaps (1978). Figure 4.2 illustrates the three different regions that have been used for this parameterization:

- One for $60°$ N - $60°$ S and from 18 - 30 km (1)
- One for $60°$ N - $60°$ S and altitudes higher than 30 km (2)
- One for geomagnetic latitudes bigger than $60°$, i.e. for the polar caps (3)

4.2 Parameterizations

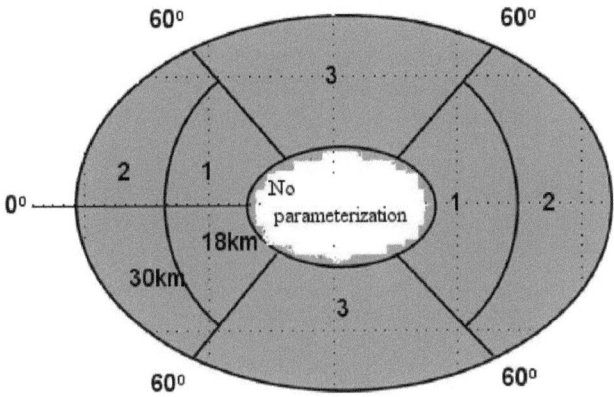

Figure 4.2: *Overview of the different regions covered by Heaps parameterization.*

The ion-pair production rate can be parameterized from $60°$ N - $60°$ S through the 18 - 30 km altitude range by

$$Q = (A + B \cdot \sin^4 \Lambda) \cdot N_0^\gamma \cdot N^n (cm^{-3} \, s^{-1}) \quad (4.1)$$

where A and B are parameters, Λ is the magnetic latitude, N is the total number density (molecules cm^{-3}), n=0.6+0.8·cosΛ, γ=1-n and N_0 is a reference number density, chosen as $N_0 = 3.03 \cdot 10^{17}$ molecules cm^{-3} at about 31 km. Table 4.1 gives the values for the parameters A and B. Note that B varies during the solar sunspot cycle.

For altitudes above approximately 30 km and from $60°$ N - $60°$ S the ion-pair production rate may be expressed as

$$Q = (A + B \cdot \sin^4 \Lambda) \cdot N (cm^{-3} \, s^{-1}) \quad (4.2)$$

Model description, parameterizations and experimental setup

	A	B (solarmax)	B (solarmin)
N(mol. cm^{-3})	$1.74 \cdot 10^{-18}$	$1.93 \cdot 10^{-17}$	$2.84 \cdot 10^{-17}$
p (Nm^{-2})	$4.6 \cdot 10^{-4}$	$5.10 \cdot 10^{-3}$	$7.50 \cdot 10^{-3}$

Table 4.1: *Overview of the Parameters A and B.*

The ionization rate over the polar caps can best be parametrized by the form

$$Q = (C + D \cdot \cos E) \cdot N (\text{cm}^{-3} \text{ s}^{-1}) \quad (4.3)$$

Where C and D are listed Table 4.2, N is the number density (pressure may also be used) and E is the argument of the cosine.

$$E = 2 \cdot \pi \cdot (YR - YM)/P \quad (4.4)$$

YR is the year of interest centered about YM; YM is the year of the cosmic ray maximum, taken as mid-year 1954, 1965 and nominally 1976; P is the periodicity of the solar cycle.

	C	D
N(mol. cm^{-3})	$1.44 \cdot 10^{-17}$	$4.92 \cdot 10^{-18}$
p (Nm^{-2})	$3.8 \cdot 10^{-3}$	$1.3 \cdot 10^{-3}$
P (the period)	=9, YR=1954 - 1958	,YM=1954
	=9, YR=1965 - 1969	,YM=1965
	=13, YR=1959 - 1965	,YM=1965
	=5, YR=1970 - 1972	,YM=1972
	=17, YR=1972 - 1980	,YM=1972

Table 4.2: *Overview of the Parameters C and D.*

With these equations (4.1 - 4.3), we are able to calculate the ionization rates. One weakness of this parameterization is that it stops at 18 km altitude even though the most intense interaction with the atmosphere happens at about 12 to 14 km which can be seen in Fig. 1.1.

4.2 Parameterizations

Usoskin parameterization

Beside the approach of Heaps, a second parameterization of the Cosmic Ray induced ionization (CRII) effect has been included into SOCOL. The second approach is based on the recently developed CRAC:CRII (Cosmic Ray induced Cascade: Application for Cosmic Ray induced ionization) model (Usoskin et al., 2004; Usoskin & Kovaltsov, 2006) which has been extended toward the upper atmosphere (Usoskin, Kovaltsov & Mironova, 2010). The model is based on a Monte-Carlo simulation of the atmospheric cascade and reproduces observed data within 10 % accuracy in the troposphere and lower stratosphere (Bazilevskaya et al., 2008; Usoskin et al., 2009). The results of the CRAC:CRII model are first interpolated to the SOCOL grid and then used (see Usoskin et al., 2005) to parameterize ion pair production rate as a function of the altitude (quantified via the barometric pressure), geomagnetic latitude (quantified via geomagnetic cutoff rigidity, see Fig. 4.3) and solar activity (quantified via the modulation potential θ). The modulation potential (given in Megavolts (MV)) provides a good single-parameter approximation of the actual shape of the GCR spectrum near Earth (Usoskin et al., 2005).

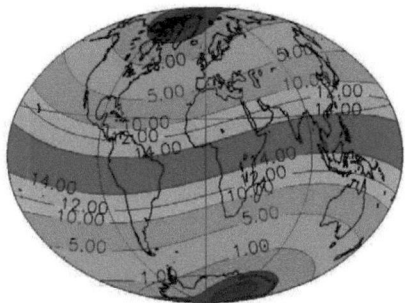

Figure 4.3: *Geomagnetic cutoff given in Gigavolts (GV). The closer to the equator, the larger the cutoff, i.e. only high energetic particles are able to penetrate deep into the equatorial atmosphere.*

Model description, parameterizations and experimental setup

4.2.2 SPE runs

Solar protons have the same restrictions like the GCRs, i.e. their ionization effect depends on altitude, latitude and solar activity. Since solar protons are less energetic than GCRs, their occurrence is limited to approximately 60° - 90° geomagnetic latitude. In contrast to the high energetic protons of the GCRs, solar protons are not able to penetrate down to the surface. The following subsections describe the parameterizations of two specific solar proton events, the Carrington event from 1859 and the October/November 2003 event. These SPEs belong to most significant events from the last 450 years.

Carrington event

The Carrington event is a special solar proton event that happened in August/September 1859. It is known as the most significant SPE for the last 450 years, about four times larger than the solar proton fluence of the largest event from the spacecraft era (Rodger et al., 2008) . Since no observational information about the Carrington event are available, data from the August 1972 event have been used to reproduce the Carrington event. The energy levels of the proton flux for the Carrington event was described by a Weibull distribution (Weibull, 1951) similar to the SPE of August 1972. The total fluence at energies > 30 MeV was scaled to match the value given by Smart et al. (2006) based on ^{10}Be isotope measurements in a polar ice core. The energy deposition and ionization rates wrt. altitude were then calculated using a method utilizing energy-range measurements for protons (Verronen et al., 2005). This approach is identical to the one used and described in more detail in Rodger et al. (2008).

4.2 Parameterizations

October/November 2003

For this solar proton event, ionization rates from two different sources have been used.

One source provided tabulated daily averaged ionization rates, valid from 1963 - 2008 and $60°$ to $90°$ as functions of pressure between 888 hPa (1 km) and $8 \cdot 10^{-5}$ hPa (115 km) at the SOLARIS (Solar Influence for SPARC) website (www.geo.fu-berlin.de/en/met/ag/strat/forschung/SOLARIS/Input_data/index.html).

Figure 4.4 shows the ionization rates taken from the SOLARIS website. It is clearly visible that the ionization took place on two different dates with the first event several times stronger.

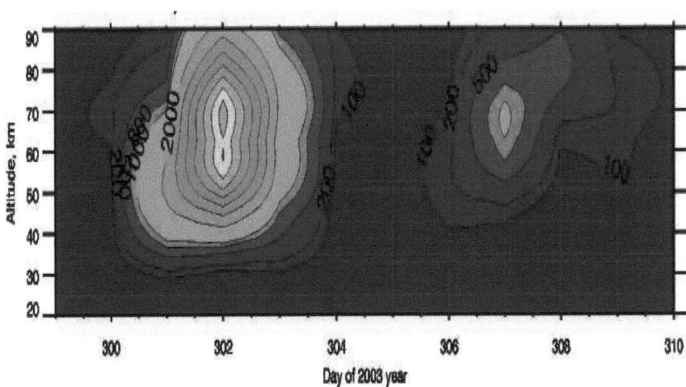

Figure 4.4: *Ionization rates given in $cm^{-3}\ s^{-1}$ starting end of October until the beginning of November provided from the SOLARIS website. The contour levels for the ionization rates are: 100, 200, 500, 800, 1000, 2000, 3000, 4000, 5000, 6000, 7000, 8000, 9000*

Model description, parameterizations and experimental setup

Figure 4.5 shows the other dataset of ionization rates that has been calculated and provided by J.M. Wissing with the ionization model AIMOS (Wissing & Kallenrode, 2009). This calculation shows the same behaviour like the IR from the SOLARIS website, even though Wissing & Kallenrode (2009) show a more intense ionization at both dates.

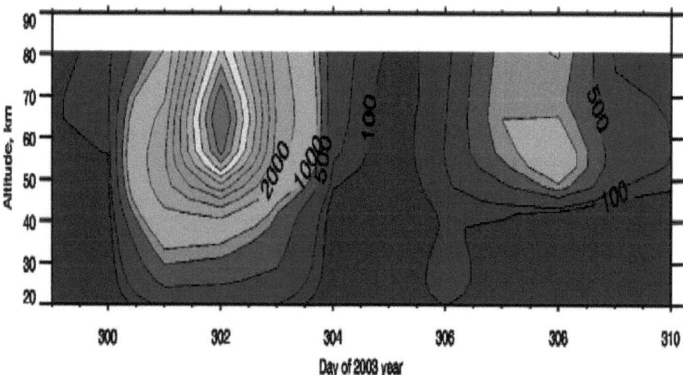

Figure 4.5: *Ionization rates given in cm^{-3} s^{-1} starting end of October until the beginning of November calculated with the ionization model AIMOS. The contour levels for the ionization rates are: 100, 200, 500, 800, 1000, 2000, 3000, 4000, 5000, 6000, 7000, 8000, 9000, 10'000*

4.2 Parameterizations

4.2.3 EEP runs

In principle, energetic electrons do have the same restrictions as GCRs and SPEs, but due to the fact that EEPs are lowest in energy, they usually lose most of their energy in an altitude range between the thermosphere and the upper part of the mesosphere. Electrons with higher energies, so-called relativistic electrons, are able to penetrate down to about 50 km.
The following subsections describe two different parameterizations for energetic electrons. The first just takes into account low energetic electrons (LEE). The second parameterization considers both, low and high energetic electrons, however, just for one year.

Low energetic electrons

The parameterization of LEE is based on the work of Baumgärtner et al. (2009). The additional NO_x production depends on a measure of magnetic activity of the Sun which can be described by the A_p index (Baumgärtner et al., 2009). Figure 4.6 shows that the flux of the odd nitrogen (see 4.5 4.6) varies in time due to the different A_p index. A peak in activity end of October is also visible in Fig. 4.6.
The equation which describes the flux for the southern hemisphere is the following.

$$F = A_p{}^{2.5} \cdot c \cdot 2.2 \cdot 10^5 cm^{-2} \cdot s^{-1}$$
$$\cdot max(0.1, cos(\pi/182.625 \cdot (d - 172.625))) \quad (4.5)$$

In order to assess the capability of the model and the parameterization in the northern hemisphere, the function for the flux is given as follows.

$$F = A_p{}^{2.5} \cdot c \cdot 2.2 \cdot 10^5 cm^{-2} \cdot s^{-1}$$
$$\cdot max(0.1, cos(\pi/182.625 \cdot (d - 355.25))) \quad (4.6)$$

where c=0.23 for average excess NO_x and d is day of the year. The sinusoidal variation centered around solstice represents the minimum requirement of a seasonal variation with maximum in winter.
A minimum absolute latitude of $55°$ has been used, i.e. low energetic electrons are not allowed to enter in the atmosphere lower than this latitude.
Because low energetic electrons lose their energy already in the thermosphere a sponge layer had to be implemented in the model, i.e. the initial ionization

Model description, parameterizations and experimental setup

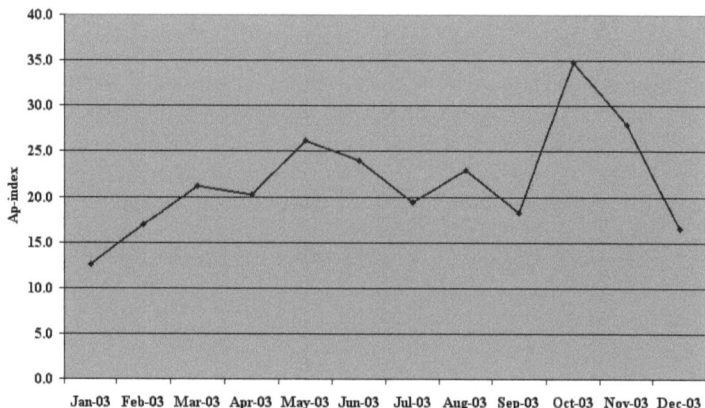

Figure 4.6: *Ap index of the Sun for 2003. Clearly visible is the maximum during October/November 2003.*

takes place in the two uppermost layers of the model. This leads to a downward transport of NO_x species below these layers. This is because the LEE are usually not able to penetrate deeper than the model top and directly ionize the air.

To differ between northern hemisphere (NH) and southern hemisphere (SH), Baumgärtner et al. (2009) used the solstice in their parameterization, i.e. depending on which hemisphere gets ionized, the day of the solstice changes.

Low energetic and high energetic electrons

This parameterization has the advantage, that the LEE and the high energetic electrons (HEE) have been taken into account. The deficiency, however, is that the ionization rates are computed for just one year, i.e for a multiyear run, the IR repeats every year. A description how the ionization rates have been calculated can be found in Rozanov et al., 2005. As seen in table 4.3 these parameterizations have been used for several runs. The outcome of these model runs are not shown in this thesis.

42

4.2 Parameterizations

4.2.4 Coupled run

Beside the above mentioned model studies seperating the different types of energetic paricles, an additional model run including all particle types has been performed. The aim of this run is to examine how the Earth's atmosphere will response when the GCRs, SPEs and EEPs are implemented simultaneously.
The following parameterizations for the different energetic particles have been taken.

- **GCR:** Concerning the Galactic cosmic rays, the parameterization by Usoskin et al. (2010) has been picked. This parameterization provides, in contrast to Heaps (1978), an ionization rate down to the surface which is more preferable than ionization rates that stop at 18 km.

- **SPE:** For the solar protons, daily ionization rates available from the SOLARIS website, valid from 60° to 90° as functions of pressure between 888 hPa (1 km) and $8*10^{-5}$ hPa (115 km) have been taken. The advantage of the SOLARIS dataset is that it provides IR from 1963 to 2008, whereas the Wissing & Kallenrode (2009) data set is limited to the SPE from October/November 2003.

- **EEP:** To consider energetic electrons, the parameterization of Baumgärtner et al. (2009) has been chosen, since it provides IR for several years and not only for just one year. However, it only takes into account low energetic electrons. Therefore, HEE have been neglected in the model run where the particles are implemented simultaneously.

Model description, parameterizations and experimental setup

4.3 Experimental setup

Table 4.1 describes the characteristics of the model runs with the different energetic particles that have been performed in this thesis.
Depending on which particle has been modeled, the runs had a different duration. Since the SPEs last just for a few days, the model runs, which had each 5 ensemble members, lasted for several months, whereas the runs with GCRs and EEPs had a duration for several years.

- **GCRs:** *Heaps:* Two different runs have been performed, a perturbed run, i.e. with the influence of the Galactic cosmic rays and an unperturbed run. The modelruns started 1976 and ended in 1988. For the comparison with the Usoskin et al. parameterization, we performed another run with the Heaps parameterization that started in 1976 and ended in 2002. To get rid of eventual spin up problems, the first two years were neglected, i.e. at the end we used datasets from 1978 to 1988 and 1978 to 2002.
- **GCRs:** *Usoskin:* We have carried out two 27-year long runs from 1976 to 2002. The control run has been performed without the influence of the Galactic cosmic rays, while the experiment run includes GCRs using the IR given by Usoskin et al., (2010) up to 0.01 hPa. The first two years of the runs have been neglected for the analysis to avoid possible spin up problems of the model.

- **SPEs:** *Carrington event:* Again, two model runs have been performed, one control run without the solar proton event and one experiment run simulating the Carringto event. Each of the runs has 5 ensemble members. The runs start in August 1972 and end in November of the same year. The reason to use ensemble members is, that for such a short time span the natural variability can be better catched with such a method. For the data analysis, the mean of the ensemble members has been taken.
- **SPEs:** *October/November 2003:* For each of the given datasets two seperate model runs have been performed with each 5 ensemble members, one with the influence of the solar protons and one without. The runs start on 1 October 2003 and end on 30 November 2003. At the end, the mean of the ensemble members has been taken for the data analysis.

- **LEEs:** Several model runs with the Baumgärtner et al. (2009) parameterization have been performed. One just for the NH, another just for

4.3 Experimental setup

the SH, and the third included both, NH and SH. For all the experiments, two different runs were made, one with the influence of the LEEs and the other without. All runs start in 1976, but the runs that distinguish between NH and SH end in 1988. The run that coupled the NH and the SH ends in 2002. To analyze the data from the different experiments, the first two years have been neglected to get rid of eventual spin up problems.

- **LEEs & HEEs:** Three different model runs, one with just the LEE, another with the HEE and the third run including both LEE and HEE have been performed. For all the experiments, two different runs were made, one with the influence of the energetic particles and another without the influence of the energetic electrons. All the runs cover the period 1976 to 2002. Again, the first two model years have been neglected to avoid spin up problems.

- **Coupled run:** Two different runs have been performed, a perturbed run, i.e. with the influence of all particles together and a reference run, without any influence of the GCRs, SPEs and EEPs. The model runs start in 1960 and end in 2005. To get rid of eventual spin up problems, the first two years were neglected, i.e. at the end, model results for the years 1962 to 2005 have been analyzed.

Model description, parameterizations and experimental setup

	Scenario	Model years	Description
GCRs	Heaps	1976 - 2005	transient run
	Usoskin	1976 - 2005	transient run
SPEs	Halloween 2003	Oct - Nov 2003	SOLARIS IR, 5 ensemble members
		Oct - Nov 2003	Wissing & Kallenrode IR, 5 ensemble members
	Carrington event	Aug - Nov 1972	5 ensemble members
LEE	Baumgärtner	1976 - 1988	transient run for NH & SH
		1976 - 2002	transient run NH & SH coupled
LEE, HEE	Callis	1976 - 1988	transient run for LEE & HEE
		1976 - 2002	transient run LEE & HEE coupled
Coupled	Usoskin, SOLARIS and Baumgärtner	1960 - 2005	transient run

Table 4.3: *Overview of the model experiments.*

Chapter 5

GCR results

Influence of Galactic cosmic rays on the atmospheric composition and temperature

(Published in: Atmos. Chem. Phys. Discuss., 11, 653-679, 2011)

M. Calisto[1], I. Usoskin[2], E. Rozanov[1,3], T. Peter[1]

Abstract

This study investigates the inuence of the galactic cosmic rays (GCRs) on the atmospheric composition, temperature and dynamics by means of the 3-D Chemistry Climate Model (CCM) SOCOL v2.0. Ionization rates were parameterized according to CRAC:CRII (Cosmic Ray induced Cascade: Application for Cosmic Ray Induced Ionization), a detailed state-of-the-art model describing the eects of GCRs in the entire altitude range of the CCM from 080 km. We nd statistically signicant eects of GCRs on tropospheric and stratospheric NO_x, HO_x, ozone, temperature and zonal wind, whereas NO_x, HO_x and ozone are annually averaged and the temperature and the zonal wind are monthly averaged. In the Southern Hemisphere, the model suggests the GCR-induced NO_x increase to exceed 10 % in the tropopause region (peaking with 20 % at the pole), whereas HO_x is showing a decrease of about 3 % caused by enhanced conversion into HNO_3 . As a consequence, ozone is increasing by up to 3 % in the relatively unpolluted southern troposphere, where its production is sensitive to additional NO_x from GCRs. Conversely,

[1]Institute for Atmospheric and Climate Science ETH, Zurich, Switzerland
[2]Sodankylä Geophysical Observatory, University of Oulu, 90014 Oulu, Finland
[3]Physical-Meteorological Observatory/World Radiation Center, Davos, Switzerland

in the northern polar lower stratosphere, GCRs are found to decrease O_3 by up to 3 %, caused by the additional heterogeneous chlorine activation via $ClONO_2$ + HCl following GCR-induced production of $ClONO_2$. There is an apparent GCR-induced acceleration of the zonal wind of up to 5 m/s in the Northern Hemisphere below 40 km in February, and a deceleration at higher altitudes with peak values of 3 m/s around 70 km altitude. The model also indenties GCR-induced changes in the surface air, with warming in the eastern part of Europe and in Russia (up to 2.25 K for March values) and cooling in Siberia and Greenland (by almost 2 K). We show that these surface temperature changes develop even when the GCR-induced ionization is taken into account only above 18 km, suggesting that the stratospherically driven strengthening of the polar night jet extends all the way down to the Earths surface.

Introduction

Galactic cosmic rays (GCRs) are energetic particles (mostly protons and α-particles) which originate from outside of the solar system. While their flux outside the solar system can be regarded as roughly isotropic and time-independent (at least on the time scales studied here), the intensity of GCRs near the Earth varies as a result of the modulation inside the heliosphere, i.e. the spatial region of about 200 EarthSun distances controlled by the solar wind and magnetic field. Figure 5.1 shows this modulation of GCR flux by the solar magnetic activity, with GCR intensities higher/lower during solar minimum/maximum. Variations of the cosmic ray flux depend also on particle energy: the flux of less energetic (< 1 GeV) particles vary by an order of magnitude during the solar cycle, while energetic GCRs (above 100 GeV) are hardly modulated.

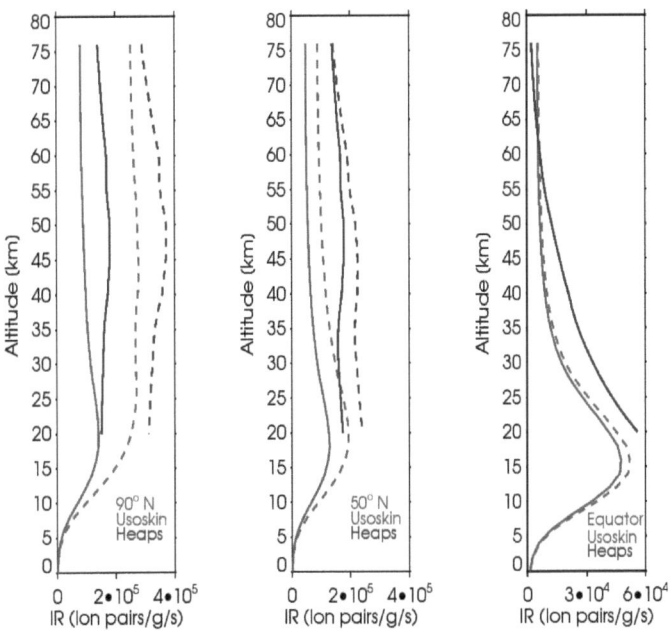

Figure 5.1: *Ionization rates (IR, number of ion pairs produced per air mass and time unit) for several geomagnetic latitudes as computed by the CRAC:CRII model (Usoskin, Kovaltsov & Mironova, 2010). Solid lines show ionization rates during solar maximum, the dashed lines during solar minimum. Note different scales on abscissas in dependence on geomagnetic latitude.*

When galactic cosmic rays enter the Earth's atmosphere they collide with the ambient atmospheric gas molecules, thereby ionizing them. In this process they may produce secondary particles, which can be suciently energetic to contribute themselves to further ionization of the neutral gases. This leads to the development of an ionization cascade (or shower). The intensity and penetration depth of the cascade depends on the energy of the primary cosmic particles. Cascades of particles with several hundred MeV of kinetic energy may reach the ground. However, due to their charge cosmic ray particles are additionally deected by the geomagnetic eld. Almost all particles can penetrate into the polar region, where the magnetic eld lines are perpendicular to the ground, whereas only the rare most highly energetic particles with energies above 15 GeV are able to penetrate the lower atmosphere near the equator.

Early models of the cosmic ray induced ionization (CRII) were (semi)empirical (e.g., OBrien, 1970; Heaps, 1978) or simplified analytical (Vitt & Jackmann; 1996, OBrien, 2005). Nicolet (1975), however, has used data from balloon soundings and ionization chambers to deduce the production rates of nitric oxide in the auroral region. State-of-the-art models (Usoskin et al., 2004; Desorgher et al., 2005; Usoskin & Kovaltsov, 2006) are based on Monte-Carlo simulations of the atmospheric cascade and can provide 3-D time dependent computations of the CRII.

Between the surface and 25-30 km CRII is the main source of the atmospheric ionization (Bazilevskaya et al. 2008) with the maximum ionization rate due to the Bragg peak in the stopping power around 15 km (Pfotzer maximum), which is clearly visible in Fig. 5.1.

The CRII leads to the production of odd nitrogen. For example, fast secondary electrons (e*) can dissociate the nitrogen molecule, $N_2 + e^* \rightarrow 2 N(^2D) + e$, and almost all of the N atoms in the ^2D state react with O_2, producing nitric oxide, $N(^2D) + O_2 \rightarrow NO + O$. CRII is estimated to lead to the production of 3.0 to $3.7 \cdot 10^{33}$ molecules of odd nitrogen per year in the global stratosphere, which amounts to about 10 % of the NO_x production following N_2O oxidation (Vitt & Jackman, 1996). In comparison, the northern polar/subpolar region ($> 50^o$ N) is believed to be supplied with NO_x in equal amounts by GCRs from (7.1 to $9.6 \cdot 10^{32}$ molecules/yr) and by N_2O oxidation (9.4 to $10.7 \cdot 10^{32}$ molecules/yr). In the deep polar winter stratosphere, when air masses experience sunlit periods only infrequently and photolysis of HNO_3 becomes negligible, CRII become the only source of NO_x, revealing the importance of GCRs in high latitudes.

Below the mesopause, where water cluster ions can be formed, CRII contributes to the formation of HO_x radicals. For example, molecular oxygen ions (O_2^+) produced by GCRs can via attachment of molecular oxygen form O_4^+, which reacts with water $O_4^+ + H_2O \rightarrow O_2^+ \cdot H_2O + O_2$. This hydrated ion quickly hydrates further to produce OH: $O_2^+H_2O + H_2O \rightarrow H_3O^+ \cdot OH + O_2 \rightarrow H_3O^+ + OH + O_2$. GCR-driven HO_x production competes with the most important source for HO_x in the atmosphere, which is the photolytically driven oxidation of water vapor (H_2O) by excited oxygen atoms, $O(^1D)$, which are themselves produced from ozone photolysis. However, during polar night, HO_x is mainly produced by the GCRs given that only little UV radiation is available for $O(^1D)$ production.

The influence of GCRs on atmospheric chemistry has been studied by Krivolutsky et al. (2002) with a 1-D photochemical model. They suggested that ozone at $50°$ geomagnetic latitude might indeed be sensitive to the additional NO_x source. Their 1-D model predicted maximum GCR-induced increases in NO_x of 4.5 % around 10 km, enhancing tropospheric ozone by 0.6 %, whereas above about 18 km ozone decreases with a maximum reduction of 0.5 % close to 20 km. Above 35 km altitude no influence caused by the GCRs was found. The evaluation of GCR-induced changes in the atmospheric temperature and dynamics, in addition to chemical changes, requires however the use of a 3-D chemistry-climate model (CCM) that is capable of describing the coupling between physicochemical processes and large-scale dynamics, which is neglected in 1-D or 2-D models.

Here we study the effect of CRII using the recently developed CRAC:CRII (Cosmic Ray induced Cascade: Application for Cosmic Ray Induced Ionization) model, and then use the results of this event-based local model to force the global CCM SOCOL, focusing on the estimation of the sensitivity of chemistry, temperature and dynamics on the influence of CRII, NO_x and HO_x from the ground to 0.01 hPa barometric pressure (altitude of approximately 80 km).

We have also addressed the difference between the state-of-the-art parameterization of the ionization rate by Usoskin et al. (2010) and the more traditional parameterization given by Heaps (1978), which was based on fitting results from sealed ionization chambers own continuously (yearly) on balloons extending to heights of 35 km. The parameterization by Heaps (1978) was and is widely used in various models (Verronen et al., 2002; Schmidt et al., 2006; Winkler et al., 2009).

Many studies of atmospheric chemistry and dynamics omit the inuence of GCRs altogether, as was done for example in the first Chemistry-Climate

Model Validation Activity (CCMVal) for coupled CCMs (Eyring et al., 2006) and in the most recent CCMValreport (see the homepage of SPARC: http://www.atmosp.physics.utoronto.ca/SPARC/ccmvalnal/index.php). Here, we use the CCM SOCOL, which is one of the CCMs that participated in the CCMVal activity, to investigate the consequences of this omission.

The models and experimental setup are described in Sect. 2, the results containing the GCR effects on several chemical species and the comparison between the parameterizations by Usoskin et al. (2010) and Heaps (1978) are presented in Sect. 3. In Sect. 4 we give a short summary of the results.

Description of the Model and experimental setup

Chemistry-Climate Modeling: The CCM SOCOL represents a combination of the global circulation model MA-ECHAM4 and the chemistry-transport model MEZON. MA-ECHAM4 (Manzini et al., 1997) is a spectral model with T30 horizontal truncation resulting in a grid spacing of about $3.75°$; in the vertical direction the model has 39 levels in a hybrid sigma-pressure coordinate system spanning the model atmosphere from the surface to 0.01 hPa.

The chemical-transport part MEZON (Rozanov et al., 1999; Egorova et al., 2003) has the same vertical and horizontal resolution and treats 41 chemical species of the oxygen, hydrogen, nitrogen, carbon, chlorine and bromine groups, which are coupled by 140 gas-phase reactions, 46 photolysis reactions and 16 heterogeneous reactions in/on aqueous sulfuric acid aerosols, water ice and nitric acid trihydrate (NAT). The original version of the CCM SOCOL was described by Egorova et al. (2005).

An extensive evaluation of the CCM SOCOL (Egorova et al., 2005; Eyring et al., 2006, 2007) revealed model deficiencies in the chemical-transport part and led to the development of the CCM SOCOL v2.0 (which is applied in this study). The new features of the SOCOLv2.0 are: (i) all species are transported separately; (ii) the mass fixer correction after each semi-Lagrangian transport step is calculated for the chlorine, bromine and nitrogen families instead for individual family members, but then applied to each individual species; (iii) the mass fixer is applied to ozone only over the latitude band $40°$ S - $40°$ N to avoid artificial mass loss in the polar areas; (iv) the water vapor removal by the highest ice clouds (between 100 hPa and the tropical cold point tropopause) is explicitly taken into account to prevent an overestimation of stratospheric water content; (v) the list of ozone-depleting substances is extended to 15 for the chemical treatment, while for the transport they are still clustered into three tracer groups; (vi) the heterogeneous chemistry module was updated to include HNO_3 uptake by aqueous sulfuric acid aerosols, a parameterization of the liquid-phase reactive uptake coefficients and the NAT particle number densities are limited by an upper boundary of $5 \cdot 10^{-4}$ cm^{-3} to take account of the fact that observed NAT clouds are often strongly supersaturated. A comprehensive description and evaluation of the CCM SOCOL v2.0 is presented by Schraner et al. (2008).

Cosmic Ray Induced Ionization Modeling: Here we study the effect of CRII using the recently developed CRAC:CRII model (Cosmic Ray induced Cascade: Application for Cosmic Ray Induced Ionization, see Usoskin et al., 2004; Usoskin & Kovaltsov, 2006) extended toward the upper atmosphere (Usoskin, Kovaltsov & Mironova, 2010). The model is based on a Monte-Carlo simulation of the atmospheric cascade and reproduces the observed data within 10 % accuracy in the troposphere and lower stratosphere (Bazilevskaya et al., 2008; Usoskin et al., 2009). In the mesosphere the agreement between observed and simulated ionizations rates are less good because the ionization by other sources (solar radiation, precipitating soft particles of magnetospheric origin, etc.) becomes at least as important as by GCRs. The results of the CRAC:CRII model are parameterized to give ion pair production rate as a function of the altitude (quantified via the barometric pressure), geomagnetic latitude (quantified via geomagnetic cutoff rigidity) and solar activity (quantified via the modulation potential θ), see Usoskin et al. (2005). In Fig. 5.1 we show the ionization rates for several geomagnetic latitudes as computed by the CRAC:CRII model (red line), compared to the ionization rates computed by the parameterization of Heaps (1978) (blue line). Solid lines show the ionization rates during solar minimum, the dashed lines during solar maximum.

This parameterization cannot be directly used in CCM SOCOL, which has no explicit treatment of ion chemistry and requires the conversion of the ionization rates into NO_x and HO_x production rates. Following Porter et al. (1976), 1.25 NO_x molecules are produced per ion pair, and 45 % of this NO_x production is assumed to yield ground state atomic nitrogen $N(^4S)$, while 55 % is assumed to go into the electronically excited. While the ground state may lower the overall NO_x concentration via $N(^4S) + NO \rightarrow N_2 + O$, $N(^2D)$ converts instantaneously to NO (see Introduction).

The production of HO_x has been studied by Solomon and Crutzen (1981) with a 1-D time-dependent model of neutral and ion chemistry. They parameterized the number of odd hydrogen particles produced per ion pair as a function of altitude and ionization for daytime, polar summer conditions of temperature, air density and solar zenith angle. We implement their parameterization in the CCM SOCOL to take into account the GCR induced production of HO_x from the ground up to the height of 0.01 hPa barometric pressure (altitude of approximately 80 km).

For this study, we have carried out three 27-yr long runs of CCM SOCOL v2.0 from 1976 to 2002. The control run has been performed without the influence of the galactic cosmic rays, while two experiment runs include GCRs using the ionization rates given by Usoskin et al. (2010) and Heaps (1978). The first two years of the runs have been omitted from the analysis to eliminate possible spin up problems of the model. In a final section we compare the results with runs using the often applied CRII parameterization of Heaps (1978).

Results

Figures 5.2-5.5 show annual mean response of the zonal mean NO_x, HO_x, HNO_3 and ozone to the GCRs calculated as a relative deviation of the experiment run from the reference run. The figures are limited to the range from 1000 hPa to 1 hPa even though the model reaches up to 0.01 hPa, because the influence of the GCRs is negligible above 1 hPa.

Chemical species:

NO_x. The galactic cosmic rays produce substantial amounts of NO_x during all seasons (not shown). In the annual mean the simulated NO_x increase affects most of the troposphere, exceeding 20 % or 10 pptv in the region extending from the south pole to 60^oN around 8-12 km altitude (see Fig. 5.2, significant changes at 95 % level are marked by hatching). There is also a significant impact on the tropical and subtropical middle stratosphere. The reason for smaller effects on the upper stratosphere lies in the vertical distribution of the ionization rates shown in Fig. 5.1: the ionization rate is the highest between 15 and 20 km, rendering the production of odd nitrogen by GCRs highest. The difference in significance between the southern to northern hemispheric troposphere is explained by the fact that more NO_x is produced anthropogenically in the Northern Hemisphere (NH) than in the Southern Hemisphere (SH), making the GCR-induced signal most relevant in the remote regions in the SH.

The annual mean NO_x production by GCRs in the southern hemispheric polar region (up to 5 pptv NO_x in January south of 70° S) is comparable to or even more important than the natural production through lightning (up to 2 pptv NO_x in January south of 70° S, Penner et al., 1998).

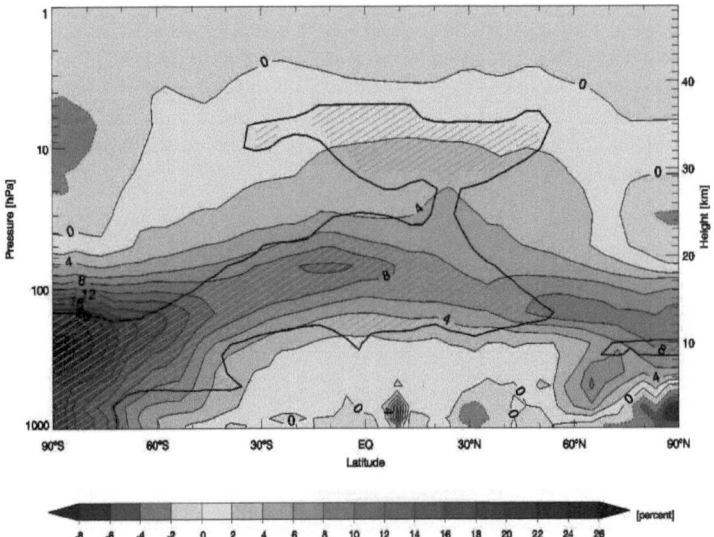

Figure 5.2: *Annual mean effect of GCRs on zonal mean NO_x, $([NO_x]_{GCR}-[NO_x]_{control})/[NO_x]_{control}$, in percent $([NO_x] = [NO] + [NO_2])$. Results are averaged from 1978-2002 (after allowing for a 2-year model spin-up) with appropriate accounting for solar minimum and maximum periods. Solid contours indicate positive, dotted contours negative changes. Hatched areas (enclosed by solid contours) indicate changes with at least 95 % statistical significance*

GCR results

HO$_x$ and HNO$_3$. Figure 5.3 represents the response of annual mean zonal mean HOx to GCRs. The GCR-induced HO$_x$ production does not result in a statistically significant HO$_x$ change in the atmosphere except a small increase in the upper tropical stratosphere and a broad area of significant GCR-induced HO$_x$ reduction in the tropical/mid-latitude UTLS. The inner hatched areas, representing 95 % statistical significance, show a decrease of about 3 % or 0.03 ppt over the southern hemispheric mid- latitudes at an altitude of approximately 20 km. The outer hatched areas, representing 80 % significance, show a decrease of up to 3 %.

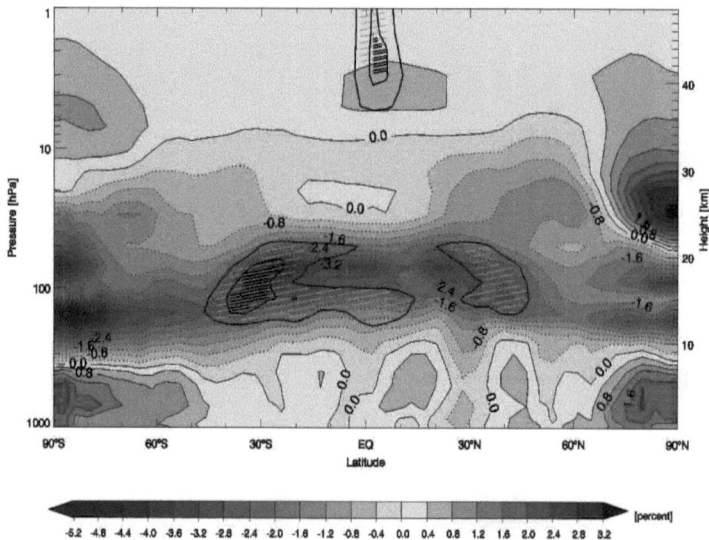

Figure 5.3: *Annual mean effect of GCRs on zonal mean HO$_x$, ([HO$_x$]$_{GCR}$-[HO$_x$]$_{control}$)/[HO$_x$]$_{control}$, in percent ([HO$_x$] = [H] + [OH] + [HO$_2$]). Results are averaged from 1978-2002 (after allowing for a 2-year model spin-up) with appropriate accounting for solar minimum and maximum periods. Solid contours indicate positive, dotted contours negative changes. Hatched areas (enclosed by solid contours) indicate statistically significant changes with at least 95 % (inner contours) or 80 % (outer contours).*

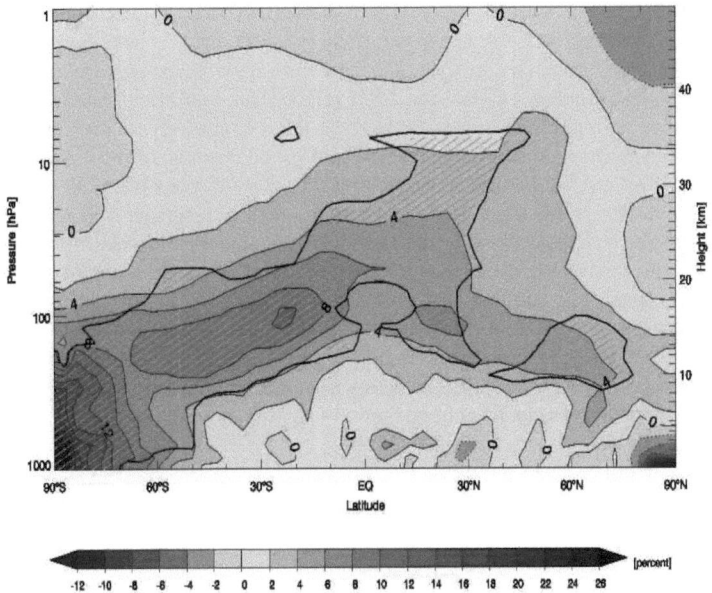

Figure 5.4: *Annual mean effect of GCRs on zonal mean HNO_3, $([HNO_3]_{GCR}-[HNO_3]_{control})/[HNO_3]_{control}$, in percent. Results are averaged from 1978-2002 (after allowing for a 2-year model spin-up) with appropriate accounting for solar minimum and maximum periods. Solid contours indicate positive, dotted contours negative changes. Hatched areas (enclosed by solid contours) indicate changes with at least 95 % statistical significance.*

This broad area of HO_x decrease coincides with a region of the highest NO_x enhancements and can be explained by the more intensive removal of OH via OH + NO_2 + M → HNO_3 + M, resulting in a significant HNO_3 increase of about 8 % (Fig. 5.4).

Ozone. Significant increase of NO_x in the southern hemispheric troposphere leads to the statistically significant ozone enhancement. As mentioned above ozone photochemistry in the southern hemispheric troposphere is in large parts NO_x-limited, so that the CRII relaxes this limitation leading to up to 3 % or 1 ppb ozone increase (see Fig 5.5). Conversely, in the northern polar lowermost stratosphere a significant ozone decrease of more than 3 % is caused by the additional production of $ClONO_2$ via $ClO + NO_2 + M \rightarrow ClONO_2 + M$, which in a second step reacts in heterogeneous reactions on polar stratospheric cloud particles or cold sulphate aerosols to enhance chlorine activation, $ClONO_2 + HCl \rightarrow Cl_2 + HNO_3$ (with subsequent Cl_2 photolysis). In addition, higher HNO_3 concentration in the polar winter stratosphere leads to enhanced polar stratospheric cloud occurrences, and hence to faster heterogeneous chemical processing. The ozone decrease by the activated chlorine is negligible in the southern hemispheric polar region because the background concentration of chlorine is too high.

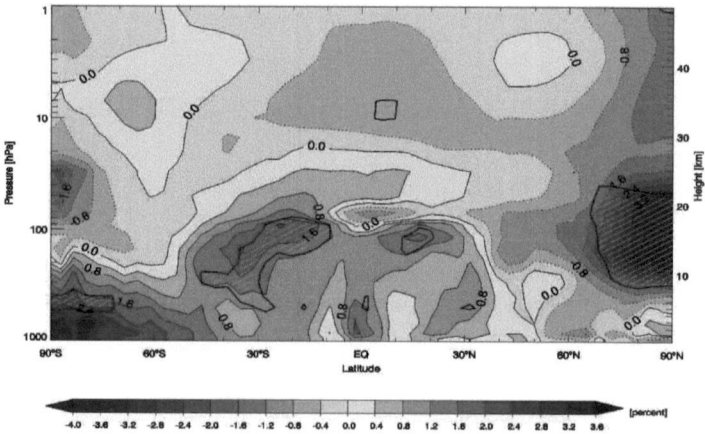

Figure 5.5: *Annual mean effect of GCRs on zonal mean ozone, $([O_3]_{GCR}-[O_3]_{control})/[O_3]_{control}$, given in percent. Results are averaged from 1978-2002 (after allowing for a 2-year model spin-up) with appropriate accounting for solar minimum and maximum periods. Solid contours indicate positive, dotted contours negative changes. Hatched areas (enclosed by solid contours) indicate changes with at least 95 % statistical significance.*

The latitudinal average of our results for O_3 and NO_x resemble the results of simple 1-D model calculations published by Krivolutsky et al. (2002). For ozone they modeled a maximum increase in the troposphere at a height of approximately 10 km and a maximum decrease at about 20 km. For NO and NO_2, their peak is visible at 10 km. In their work the inuence of the GCRs vanishes above 35 km. The hemispheric asymmetries discussed above could, of course, not be retrieved in their 1-D calculation. Also, because Krivolutsky et al. (2002) did not discuss HO_x in their paper, it is not possible to make a quantitative comparison with our results.

Temperature and winds

The effects fo GCRs on monthly mean ozone, temperatures and winds are noticeable year-round. However, signficance is highest for winter/spring, hence results are displayed for this season in Fig. 5.6. The upper panel in Fig. 5.6 shows the monthly mean zonal mean changes for ozone during February. The significant area and the percentage of decrease are similar to the annual mean results shown in Fig 5.5. A decrease of up to 5 % or more than 60 ppbv is visible in the NH polar region between 20 and almost 30 km. The influence of the GCRs on the SH is strongest in the troposphere, but remains statistically insignificant on the 95 % level. As discussed above, the reason for the ozone depletion in the NH polar region is the additional GCR-induced chlorine activation

Temperature profile: The center panel of Fig. 5.6 shows zonal mean response of the temperature in February. There is a cooling in the NH lowermost stratosphere (below 20 km altitude), resulting from the radiative cooling caused by the ozone loss. This cooling is facing a warming at low latitudes at low altitudes at about $40°$ N. These two effects lead to an increase in the latitudinal temperature gradient in the lowermost stratosphere. In addition, there is a significant warming between 40 and 50 km in the NH polar region due to an intensification of the polar vortex which leads in turn to the increase of air descent and adiabatic warming of the upper stratosphere.

Zonal wind profile: The influence on the monthly mean zonal wind for February shows a significant increase of up to 5 m/s in the NH polar region, peaking in the lower stratosphere and extending all the way to the ground (see lower panel in Fig. 5.6). The acceleration is caused by the cooling of the polar lower stratosphere due to the GCR-induced polar ozone depletion, opposed to the warming of the northern mid-latitude lowermost stratosphere. These changes increase the meridional temperature gradient, leading to acceleration of the zonal wind in agreement with the thermal wind balance. Intensification of the polar vortex leads in turn to the increase of air descent and adiabatic warming of the upper stratosphere, in turn causing deceleration of the zonal wind (Limpasuvan et al., 2005).

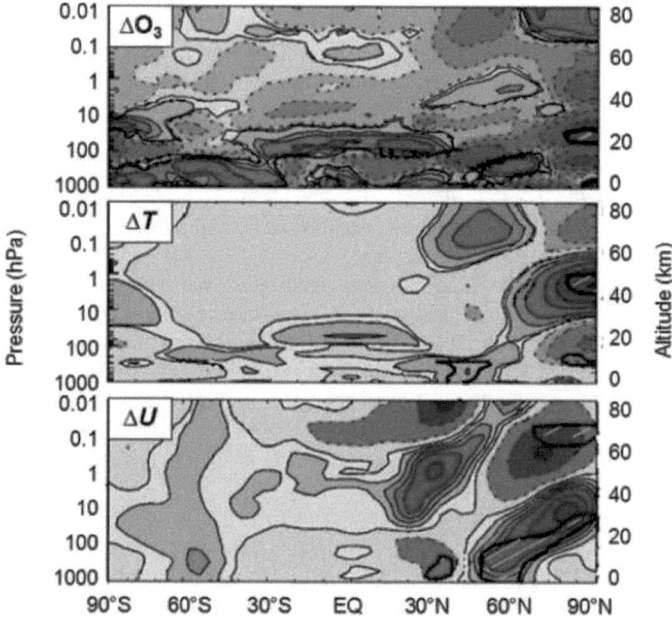

Figure 5.6: *Monthly mean zonal mean effects of GCRs on ozone (O_3), temperature (T) and zonal wind (U) for the month of February. Upper panel: effect on O_3 given in percent (in steps of 0.5 %). Center panel: effect on T given in Kelvin (in steps of 0.25 K). Lower panel: effect on U given in m/s (in steps of 0.5 m/s). Hatched areas show 95 % statistical significance.*

Comparison of the Heaps and CRII parameterizations. As mentioned above, a major difference between the Heaps parameterization and Usoskin's model-based approach is that the ionization rate calculated with the Heaps parameterization is applicable only at altitudes above 18 km, whereas the ionization rates derived by Usoskin extend to the ground. As described above, a proper description of the ionization rate in the upper troposphere and lower stratosphere is required for a correct simulation in of atmospheric composition, in particular of free tropospheric ozone.

The importance of the accuracy of the GCR parameterizations for ozone is illustrated in Fig. 5.7. The left panel represents the annual mean effect of GCRs on the zonal average ozone at 70° - 90° N given in percent averaged from 1978 to 2002. It reveals that the Heaps parameterization (Heaps, blue line; Usoskin, red line) clearly underestimates the ozone decrease due to the additional NO_x that is produced in the UTLS region, because it neglects the ionization below 18 km altitude. Conversely, the right panel of Fig. 5.7 shows the importance of the Usoskin scheme for correctly describing the GCR-induced ozone production in the southern hemispheric troposphere.

The upper panel in Fig. 5.8 shows the monthly mean zonal mean effect of the GCRs on ozone at 70° - 90° N for November and December given in percent whereas the lower panel depicts the changes for February and March. The larger and significant decrease in November at about 30 km with Heaps parameterization is caused through the fact that the ionization rate is larger in the middle stratosphere (see Fig. 5.1) and that the PCS chemistry is not important yet. This changes in February and March: the lower panel in Fig. 5.8 shows that the parameterization with Usoskin below altitudes of 20 km shows a larger impact on ozone than the Heaps parameterization which stops at 18 km.

Finally, we investigate the inuence of the two GCR parameterizations on the surface air temperature (SAT) and its connection with the Arctic Oscillation. The upper panels of Fig. 5.9 show the March monthly mean and the annually averaged changes in SAT for ionization rates calculated by Usoskin et al. (2010), and the lower panels show the January monthly mean and the annually averaged changes but using the parameterization by Heaps (1978). The general patterns and intensities for both parameterization are in good agreement: both show a warming over the eastern part of Europe and Russia and a cooling in the high Arctic.

Additionally, we see that the effects of the galactic cosmic rays result in an alternating warming/cooling pattern resembling the typical response of the SAT caused by an intensication of the polar vortex known as positive

GCR results

phase of Arctic Oscillation (Thompson and Wallace, 1998), termed AO$^+$. A resulting interesting question is whether this response is primarily due to GCR-induced stratospheric changes or due to the penetration of the GCRs into the troposphere. The presence of the AO$^+$-like warming-cooling pattern also for the Heaps parameterization, which ignores GCR-effects below 18 km, corroborates the interpretation of Thompson and Wallace (1998), namely 'that under certain conditions, dynamical processes at stratospheric levels can affect the strength of the polar vortex all the way down to the earths surface'.

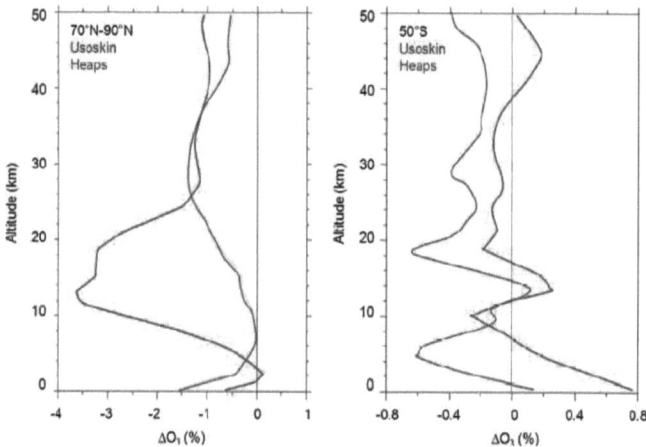

Figure 5.7: *GCR-induced effects on ozone, $([O_3]_{GCR}-[O_3]_{control})/[O_3]_{control}$, given in percent. Left panel shows the annual mean averaged for 70° - 90° N and for 50° S (right). Red line: parameterization by Usoskin et al. (2010). Blue line: parameterization by Heaps (1976). Results are averaged from 1978-2002 (all seasons, after allowing for a 2-yr model spin-up)*

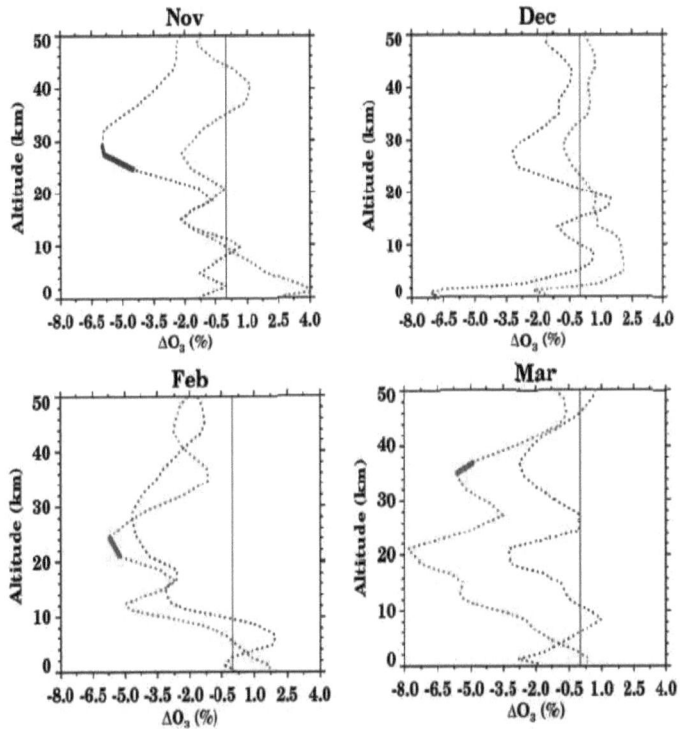

Figure 5.8: *GCR-induced effects on ozone, $([O_3]_{GCR}-[O_3]_{control})/[O_3]_{control}$, given in percent, f or 70° - 90° N. Upper panel: November and December; lower panel: February and March. Red lines: parameterization by Usoskin et al. (2010). Blue lines: parameterization by Heaps (1976). Thick solid lines: altitudes where the changes in ozone are significant at 95 % level for the respective parameterization. Results are averaged from 1978-2002.*

GCR results

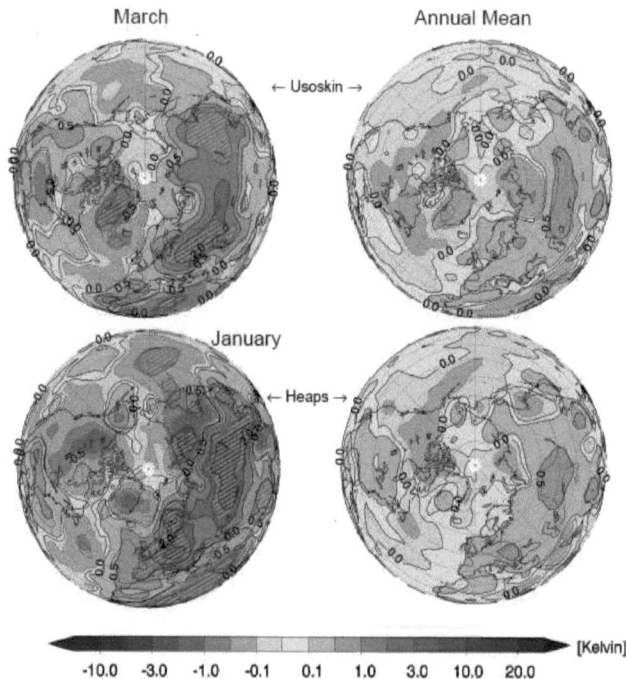

Figure 5.9: *Effect of GCRs on SAT, $[SAT]_{GCR}$-$[SAT]_{control}$, given in Kelvin for January monthly mean (left) and annual mean (right). Upper panels: using ionization rate modeled by Usoskin et al. (2010). Lower panels: using parameterization by Heaps (1978). Results are averaged from 1978-2002 (after allowing for a 2-yr model spin-up). Reddish colors: positive changes. Bluish colors: negative changes. Hatched areas (enclosed by thick solid lines) indicate changes with at least 95 % statistical significance.*

Summary

Based on the 3-D CCM SOCOL v2.0 and on CRAC:CRII (the Cosmic Ray induced Cascade: Application for Cosmic Ray Induced Ionization) model, we present in this paper a modeling study of the influence of the galactic cosmic rays on atmospheric composition, winds and temperature from 0.01 hPa or approximately 80 km down to the ground.

Our calculations indicate that GCR-induced ionization leads to the following modifications in atmospheric composition, winds (U), atmospheric temperatures (T) and surface air temperatures (SAT). Only results with 95 % level of statistical significance are stated:

Southern hemispheric troposphere, pristine conditions:

- NO_x: increases by more than 20 % in the polar region,
- HO_x: decreases of 3 % in the mid-latitude upper troposphere,
- HNO_3: increases by more than 10 % between the South Pole and subtropics,
- O_3: increases up to 3 % throughout the troposphere to 20 km between the South Pole to 20° N,
- SAT: small patches of (significant) warming up to 0.5 K in Antarctica.

Northern hemispheric troposphere, anthropogenically preconditioned:

- HNO_3: marginally significant increases in the mid-latitude upper troposphere,
- O_3: marginally significant decreases in the polar upper troposphere,
- U: enhancements of the polar night jet by up to 5 m/s at the tropopause reaching all the way to the ground,
- SAT: warming up to 2.25 K in the eastern part of Europe and Russia and decreases by almost 2 K over Greenland.

Southern hemispheric stratosphere:

- NO_x: increases up to 4 % in the tropical middle stratosphere,
- HO_x: decreases by up to 3 % caused by OH + NO_2 producing HNO_3 in the low latitude lower stratosphere,

GCR results

- HNO_3: largely mirroring the HO_x changes, with increases by 4 % in the low latitude lower stratosphere,

- U: enhancements of the polar night jet by up to 3 m/s.

Northern hemispheric stratosphere:

- NO_x: increases up to 4 % in the tropical middle stratosphere,

- HO_x and HNO_3: similar to southern hemisphere,

- O_3: strong loss in the polar lower stratosphere with annual mean mixing ratios decreasing by 3 % due to additional chlorine activation (specifically in February decreases up to 5 %, corresponding to a loss of > 60 ppb),

- T: cooling by up to -1.5 K in the lower polar stratosphere, opposed to a slight warming (< + 0.5 K) in the tropical lower stratosphere and a moderate warming (< + 1.5 K) in the upper polar stratosphere,

- U: enhancements of the polar night jet by up to 5 m/s resulting from the enhanced meridional temperature gradient in the lower stratosphere, and a decrease by 3 m/s in the mesosphere.

We conclude that for NO_x - limited regions it is important to have a parameterization for the GCRs that extents to the surface, otherwise important consequences for tropospheric ozone (Fig. 5.5) and for the oxidation capacity of the troposphere (Fig. 5.3) will be neglected. Conversely, Galactic cosmic rays appear to affect winds and temperatures in the middle and lower atmosphere in a manner that is governed by the ionization processes in the middle atmosphere alone, i.e. a detailed description of the ionization processes in the troposphere appears to be less important. The comparison between the often applied parameterization of ionization rates derived by Heaps (1978) and the state-of-the-art modeling work by Usoskin et al. (2010), which agree largely above but differ below 18 km, reveals that changes in the surface air temperature are to first order independent of the choice of parameterization. This suggests that the changes in tropospheric meteorology depend on changes in the stratosphere, i.e. that the acceleration of the polar night jet reaches all the way down to the Earth's surface. This constitutes an example of stratosphere-troposphere coupling. Conversely, tropospheric NO_x

and ozone depend strongly on a correct description of the GCRs down to the lowest parts of the troposphere. The simulations with the 3-D chemistry-climate model SOCOL show that the inuence of the GCRs should not be neglected in investigations of the tropospheric and stratospheric chemistry and dynamics.

References

Bazilevskaya, G.A., Usoskin, I.G., Flckiger, E.O., Harrison, R.G., Desorgher, L., Btikofer, R., Krainev, M.B., Makhmutov, V.S., Stozhkov, Y.I., Svirzhevskaya, A.K., Svirzhevsky, N.S. and Kovaltsov, G.A.: Cosmic Ray Induced Ion Production in the Atmosphere, Space Sci. Rev., 137, 149, 2008.

Crutzen, P.J.: The influence of nitrogen oxides on the atmospheric ozone content, Quart. J. R. Met. Soc., 96, 320 325, 1970.

Desorgher, L., Flckiger, E.O., Gurtner, M., Moser, M. and Btikofer, R.: Atmocosmics: a Geant 4 Code for Computing the Interaction of Cosmic Rays with the Earth's Atmosphere, Internat. J. Modern Phys. A, 20, 6802-6804, 2005.

Egorova, T., Rozanov, E., Zubov, V. and Karol, I.L.: Model for Investigating Ozone Trends (MEZON), Izvestiya, Atmospheric and Oceanic Physics, 39, 277-292, 2003.

Egorova, T., Rozanov, E., Zubov, V., Manzini, E., Schmutz, W., and Peter, T.: Chemistry-climate model SOCOL: a validation of the present-day climatology, Atmos. Chem. Phys., 5, 1557-1576, 2005.

Eyring, V., et al.: Assessment of temperature, trace species, and ozone in chemistry-climate model simulations of the recent past, J. Geophys. Res., 111, D22308, doi:10.1029/2006JD007327, 2006.

Eyring, V., et al.: Multimodel projections of stratospheric ozone in the 21st century, J. Geophys. Res., 112, D16303, doi:10.1029/2006JD008332, 2007.

Heaps, M.G.: Parametrization of the cosmic ray ion-pair production rate above 18 km, Planet. Space Sci., 26, 513-517, 1978.

Jackman, C.H., Frederick, J.E. and Stolarski, R.S.: Production of Odd Nitrogen in the Stratosphere and Mesosphere: An intercomparison of source strengths, J. Geophys. Res., 85, C12, 7495 7505, 1980.

Jackman, C.H., Douglass, A.R., Rood, R.B. and McPeters, R.D.: Effect of Solar Proton Events on the middle atmosphere during the past two solar cycles as computed using a two-dimensional Model, J. Geophys. Res., 95, D6, 7417 7428, 1990.

Krivolutsky, A., Bazilevskaya, G., Vyushkova, T. and Knyazeva, G.: Influence of cosmic rays on chemical composition of the atmosphere: data analysis and photochemical modeling, Phys. and Chem. of the Earth, 27, 471 476, 2002.

Manzini, E., McFarlane, N.A. and McLandress, C.: Impact of the Doppler spread parameterization on the simulation of the middle atmosphere circulation using the MA/ECHAM4 general circulation model, J. Geophys. Res. Atmos., 102, D22, 25751 25762, 1997.

Nicolet, M.: On the production of nitric oxide by cosmic rays in the mesosphere and stratosphere, Planet. Space Sci., 23, 637 649, 1975.

OBrien, K.: Calculated Cosmic Ray Ionization in the Lower Atmosphere, J. Geophys. Res., 75, 4357, 1970.

OBrien, K.: The theory of cosmic-ray and high-energy solar-particle transport in the atmosphere, in: The Natural Radiation Environment VII, Seventh International Symposium on the Natural Radiation Environment (NRE-VII), (J.P. McLaughlin, S.E. Simopoulos, and F. Steinhusler, Eds.), Elsevier, Amsterdam, pp. 29-44, 2005.

Porter, H.S., Jackman, C.H. and Green, A.E.S.: Efficiencies for production of atomic nitrogen and oxygen by relativistic proton impact in air, J. Chem. Phys., 65, No.1, 1976.

Prather, M.J.: Numerical Advection by conservation of 2nd-order moments, J. Geophys. Res. Atmos., 91, D6, 6671 6681, 1986.

Rahman, M., Cooray, V., Possnert, G. and Nyberg, J.: An experimental quantification of the NOx production efficiency of energetic alpha particles in air, J. Atmos. Sol. Terr. Phys 68, 1215 1218, 2006.

Rozanov, E., Callis, L., Schlesinger, M., Yang, F., Andronova, N. and Zubov, V.: Atmospheric response to NOy source due to energetic electron precipitation, Geophys. Res. Lett., 32, L14811, doi:10.1029/2005GL023041, 2005.

Rozanov, E., Schlesinger, M. E., Zubov, V., Yang , F. and N. Andronova, G.: The UIUC three-dimensional stratospheric chemical transport model: Description and evaluation of the simulated source gases and ozone, J. Geophys. Res., 104, 11,755-11,781, 1999.

Schraner, M., Rozanov, E., Schnadt-Poberaj, C., Kenzelmann, P., Fischer, A., Zubov, V., Luo, B.P., Hoyle, C., Egorova, T., Fueglistaler, S., Brnnimann, S., Schmutz W. and Peter, T.: Chemistry climate model SOCOL: version 2.0 with improved transport and chemistry/ microphysics schemes, Atmos. Chem. Phys., 8, 19, 5957-5974, 2008.

Shindell, D.T., Schmidt, G.A., Miller, R.L. and Rind, D.: Northern Hemisphere winter climate response to greenhouse gas, ozone, solar, and volcanic forcing, J. Geophys. Res., 106, D7, 7193 7210, 2001.

Simpson, J.A.: Elemental and isotopic composition of the galactic cosmic rays, Ann. Rev. Nucl. Part. Sci., 33, 323 381, 1983.

Solomon, S., Rusch, D.W., Gerard, J.-C., Reidt, G.C. and Crutzen, P.J.: The effect of particle precipitation events on the neutral and ion chemistry of the middle atmosphere: II. Odd Hydrogen, Planet. Space Sci., 29, No. 8, 885 892, 1981.

Swider, W. and Keneshea, T.J.: Decrease of ozone and atomic oxygen in lower mesosphere during a PCA event, Planet. Space Sci., 21, No.11, 1969 1973, 1973.

Thompson, D.W.J. and Wallace, J.M.: The Arctic Oscillation signature in the wintertime geopotential height and temperature fields, Geophys. Res. Lett., 25, No.9, 1297 1300, 1998.

Thorne, R.M.: The importance of Energetic Particle Precipitation on the Chemical Composition of the Middle Atmosphere, Pure and applied geophysics, 118, No.1, 128 151, 1980.

Usoskin I.G., Gladysheva, O.G. and Kovaltsov, G.A.: Cosmic ray-induced ionization in the atmosphere: spatial and temporal changes, J. Atmos. Sol. Terr. Phys., 66, 1791, 2004.

Usoskin I.G., Alanko-Huotari K., Kovaltsov G.A. and Mursula, K.: Heliospheric modulation of cosmic rays: Monthly reconstruction for 1951-2004, J. Geophys. Res., 110, A12108, 2005

Usoskin, I.G. and Kovaltsov, G.A.: Cosmic ray induced ionization in the atmosphere: Full modeling and practical applications, J. Geophys. Res., 111, D21206, doi:10.1029/2006JD007150, 2006.

Usoskin, I.G., Desorgher, L., Velinov, P., Storini, M., Flckiger, E.O., Btikofer, R. and Kovaltsov, G.A.: Ionization of the earth's atmosphere by solar and galactic cosmic rays, Acta Geophysica, 57, 88-101, 2009.

Usoskin, I.G., Kovaltsov, G.A. and Mironova, I.A.: Cosmic ray induced ionization model CRAC:CRII : An extension to the upper atmosphere, J. Geophys. Res., doi:10.1029/2009JD013142, 2010.

Vitt, F.M. and Jackman, C.H.: A comparison of sources of odd nitrogen production from 1974 through 1993 in the Earths middle atmosphere as calculated using a two-dimensional model, J. Geophys. Res., 101, D3, 6729 6739, 1996.

Wang, Y., DeSilva, A.W., Goldenbaum, G.C. and Dickerson, R.R.: Nitric oxide production by simulated lightning: Dependence on current, energy and pressure, J. Geophys. Res. 103, 19149 -19159, 1998.

Williamson, D.L. and Rasch, P.J.: Two-Dimensional Semi-Lagrangian Transport with shape-preserving interpolation, Monthly Weather Review, 117, 1, 102 129, 1989.

Zubov, V.A., Rozanov, E.V. and Schlesinger, M.E.: Hybrid scheme for three-dimensional advective transport, Monthly Weather Review, 127, 6, 1335 1346, 1999.

Chapter 6

SPE results

Influence of a Carrington like event on the atmospheric chemistry, temperature and dynamics

(In preparation for submission to: Atmospheric Chemistry and Physics, 2011)

M. Calisto[1], P.T. Verronen[2], E. Rozanov[1,3], T. Peter[1]

Abstract

We have modeled the atmospheric effects, especially concerning composition, dynamics and temperature following a major solar proton event similar to the Carrington Event of 1-2 September 1859. By means of the 3-D Chemistry Climate Model SOCOL v2.0 with an applied energy flux scaling for the Carrington Event based on ^{10}Be isotope measurements in the Greenland ice cores, we simulated the atmospheric effects under conditions of the 1976 atmosphere. For the ionization rates we have adopted the August 1972 solar proton event which shows a comparatively energy spectrum to the Carrington Event. We find significant influence on ozone, NO_x, HO_x, temperature and zonal wind. NO_x and ozone have in common an unusually strong and long lived response to this solar proton event, with upper mesospheric HO_x increase of more than 1000 %. Less intense but still 3-fold is the increase of NO_x in the upper stratosphere lasting until the end of November. Due to the enhancement in NO_x and HO_x, ozone reduces by up to 60-80 % in

---

[1]Institute for Atmospheric and Climate Science ETH, Zurich, Switzerland
[2]Finnish Meteorologica Institute, Helsinki, Finnland
[3]Physical-Meteorological Observatory/World Radiation Center, Davos, Switzerland

the mesosphere during the days after the event, and by up to 20-40 % in the middle stratosphere lasting for several months after the event. Total ozone reduces by up to 20 DU in the northern hemisphere and 40 DU in the southern hemisphere. Temperature shows a significant cooling of more than 3 K and zonal wind changes significantly by 3-5 m/s. In conclusion, a solar proton event similar to the Carrington Event from 1859 is expected to have a major impact on todays atmospheric composition throughout the middle atmosphere, resulting in an extraordinary long lasting decrease in the ozone content.

Introduction

A solar proton event, which typically lasts for a few days, occurs when protons emitted by the active Sun are accelerated to very high energies (up to 500 MeV) either close to the Sun's surface during a solar flare or in interplanetary space by magnetic shock waves associated with coronal mass ejections. These high energy protons are deflected when they penetrate the Earth's magnetic field, and upon penetration of the atmosphere can cause massive ionization in the ionosphere including significant production of HO_x and NO_x.

Solar protons normally have insufficient energy to deeply penetrate the Earth's magnetic field. However, during unusually strong solar flares, protons assume kinetic energies sufficiently high to penetrate deeper into the Earth's magnetosphere and ionosphere especially at the poles, where the magnetic field lines cut across pressure levels and dive deep into the atmosphere towards the surface. Energetic protons that are guided into the polar regions collide with atmospheric constituents and gradually transfer their kinetic energy into potential energy through the process of ionization of the air constituents. Figure 1 shows that the second event in the beginning of September is more pronounced than the first end of August and that the major fraction of the energy is deposited in the mesosphere (around 50-80 km in altitude). This ionization is able to produce fast secondary electrons (e*) which can dissociate the nitrogen molecule, $N_2 + e^* \rightarrow 2\ N(^2D) + e$. Almost all of the N atoms in the ^2D state react with O_2, producing nitric oxide, $N(^2D) + O_2 \rightarrow NO + O$. Below the mesopause, where water cluster ions can form, the ionization through the solar protons contributes to the formation of HO_x radicals. For example, molecular oxygen ions (O_2^+) produced by the solar proton event (SPE) can via attachment of molecular oxygen form (O_4

$^+$), which reacts with water. This hydrated ion quickly hydrates further to produce OH: $O_2\ ^+H_2O + H_2O \rightarrow H_3O^+ \cdot OH + O_2 \rightarrow H_3O^+ + OH + O_2$. The additional ionization resulting from a solar proton event causes the ionosphere to be more opaque to incoming high-frequency (HF) radio frequencies (Patterson et al., 2001). The enhancement of absorption of the background HF cosmic radio noise across the entire polar cap produced by solar protons is given the name polar cap absorption (PCA) (Bailey, 1964). Additionally, Odenwald et al. 2006 estimate that 80 satellites in low-, medium, and geostationary- Earth orbits might be disabled as a consequence of a Carrington like event with additional disruptions caused by the failure of many of the satellite navigation systems (e.g. GPS).

The influence of solar proton events and in particular of the 1859 Carrington Event, and the subsequent consequences on atmospheric chemistry has been assessed by Thomas et al. (2007) and Rodger et al. (2008). Thomas et al. (2007) were using the Goddard Space Flight Center two-dimensional atmospheric model. For the calculations, they scaled the solar proton event from 1989 to match to the flux of the 1859 SPE. The duration of the ionization applied by Thomas et al. (2007) is 2 days. They demonstrated that ozone is sensitive to the additional NO_x source in the atmosphere, such that two months after the event the ozone reduction at about 40 km still exceeds 30 % in both hemispheres.

Rodger et al. (2008) used a 1-D combined ion and neutral chemistry model, studying the effects on ionosphere and atmosphere. They showed that a Carrington-type event could result in an unusually strong and long lived O_x decrease (levels drop by approximately 40 %) in the upper stratosphere. Additionally, they demonstrated that such an event would cause disruption of HF communications, but this effect was found not to be significantly larger than those caused by other major SPEs.

In this paper, in addition to the chemical effects, we model the changes in temperature and dynamics after an SPE as strong as the Carrington Event, using a 3-D Chemistry Climate Model (CCM). The goal is to determine the significance of changes induced by such a type of SPE at all levels down to the troposphere. To this end we applied the CCM SOCOL, which describes the entire chain of physical and chemical processes including changes in atmospheric dynamics that are not resolved in 1-D or 2-D models. The model and experimental setup is described in section 2, section 3 gives the results while in section 4 the discussion and conclusion is formulated.

Model description and experimental setup

The CCM SOCOL is a combination of the GCM MA-ECHAM4 and the chemistry-transport model MEZON. MA-ECHAM4 (Manzini et al. 1997) is a spectral model with T30 horizontal truncation resulting in a grid spacing of about 3.75°; in the vertical direction the model has 39 levels in a hybrid sigma-pressure coordinate system spanning the model atmosphere from the surface to 0.01 hPa.

The chemical-transport part MEZON (Rozanov et al. 1999; Egorova et al. 2003) has the same vertical and horizontal resolution and treats 41 chemical species of the oxygen, hydrogen, nitrogen, carbon, chlorine and bromine groups, which are determined by 140 gas-phase reactions, 46 photolysis reactions and 16 heterogeneous reactions in/on aqueous sulfuric acid aerosols, water ice and nitric acid trihydrate (NAT). The original version of the CCM SOCOL was described by Egorova et al. (2005).

An extensive evaluation of the CCM SOCOL (Egorova et al. 2005, Eyring et al. 2006, 2007) revealed several model deficiencies in the chemical-transport part and led to the development of the CCM SOCOLv2.0. The new features and the evaluation of SOCOLv2.0 are presented by Schraner et al. (2008).

The spectrum for the Carrington like event was described by a Weibull distribution similar to the SPE of August 1972. Based on ^{10}Be isotope measurements in the Greenland ice cores, the Weibull distribution is scaled to reproduce the > 30 MeV proton fluence for this event (1.9 * 10^{10} cm^{-2}) (Smart et al., 2006). The energy deposition and ionization rates (IR) with respect to altitude were then calculated using a method utilizing energy-range measurements for protons (Verronen et al., 2005). This approach is identical to the one used and described in more detail by Rodger et al. (2008).

The above mentioned method cannot be directly used in the CCM SOCOL which has no explicit treatment of ion chemistry, therefore it is necessary to convert the ionization into NO$_x$ and HO$_x$ production.

Following Porter et al. (1976) 1.25 NO$_x$ molecules are produced by ion pair, 45 % of this NO$_x$ production is assumed to yield ground state atomic nitrogen while 55 % is assumed to go into N(^{2}D) with instantaneous conversion to NO. The production of HO$_x$ by proton precipitation has been taken into account using the calculations in Solomon and Crutzen (1981). They showed that two HO$_x$ species are produced below about 60 km altitude, and a smaller amount is produced above until the production becomes insignificant from approximately 85 km on.

For this study, we have implemented the influence of the August 1972 SPE which is similar to the event from September 1859 in the 3-D CCM SOCOL v2.0 and have carried out two 4-month long runs with 5 ensemble members each starting in August 1976 and ending in November 1976 to simulate the influence on an atmosphere with nearly today's composition. The 5 control runs and the 5 perturbed runs covered the whole 4 month period.

Results

Figures 6.2 to 6.6 show time series of zonal mean changes from 70°-90° of HO_x, NO_x, ozone, temperature and zonal wind due to a Carrington like event. Figure 6.7 shows the changes in total ozone for the NH and the SH from September to November.

The results show, that the NO_x enhancement through a Carrington like event lasts for weeks after the event has happened due to the longevity of this species. The simulated NO_x increase is higher than 300 ppbv in the SH, whereas in the NH it reaches about 200 ppbv (Fig. 6.2).

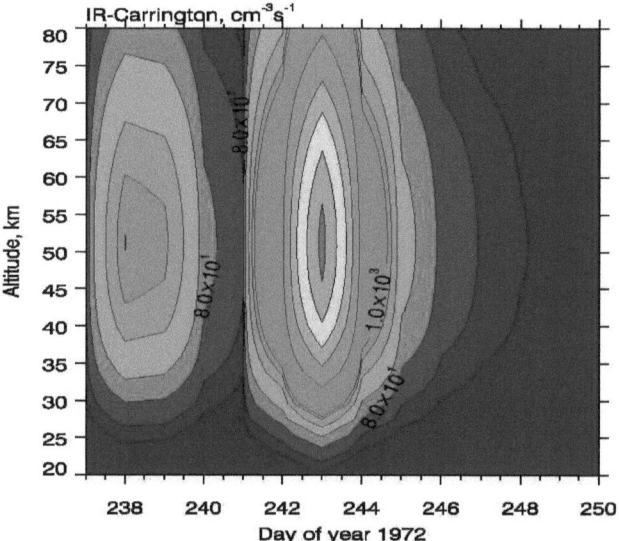

Figure 6.1: *Time altitude profile of the ionization rates from the Carrington Event given in cm^{-3} s^{-1}. Contour levels: 0, 5, 20, 80, 200, 500, 800, 1'000, 4'000, 8'000, 12'000, 18'000, 25'000, 30'000.*

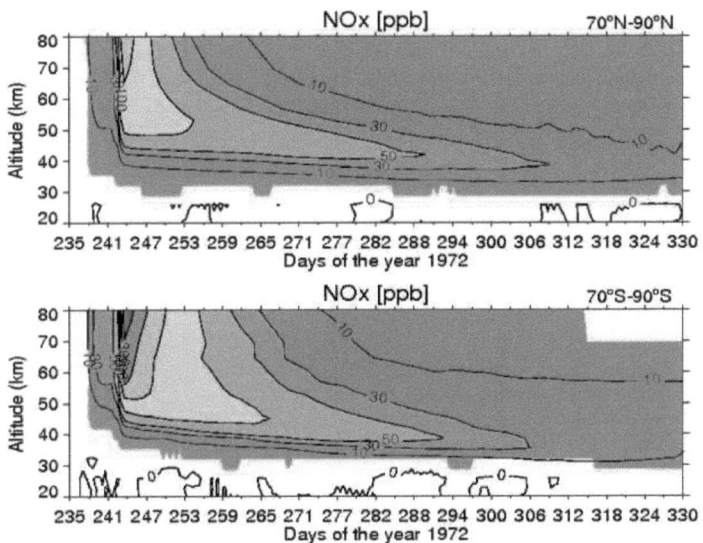

Figure 6.2: *Upper panel: Time altitude profile of NO_x zonal mean from 70° - 90° N given in ppbv. Lower panel: same for 70° - 90° S. Colores indicate areas with at least 95 % statistical significance. Contour levels (ppbv): 10, 30, 50, 100, 200, 300, 400.*

The southern hemisphere displays a stronger NO_x perturbation because the event happened on 1-2 September. This is still 3 weeks before equinox and thus the NH received the somewhat stronger exposure to this event. However, more importantly, this date still falls into the southward phase of the Brewer-Dobson circulation, i.e. is typically about 6 weeks before the first noticeable reversal of the mean meridional circulation into northward directions. Hence, most of the blast is carried southward, and here the biggest effects are observed. This result do not agree with the 1-D model used in Rodger et al. 2008 because their model has no dynamics implemented, whereas the 2-D model used in Thomas et al. 2007 is in good qualitative agreement.

SPE results

The upper panel in Fig. 6.3 representing the northern hemisphere shows an increase for HO_x of up to 10 ppbv or more than 200 % from 70 km to model top. A higher increase of more than 15 ppbv or more than 1000 % from 70 km to model top is computed for the southern hemisphere. The reason for this hemispheric difference si again the still pervailing southward transport.

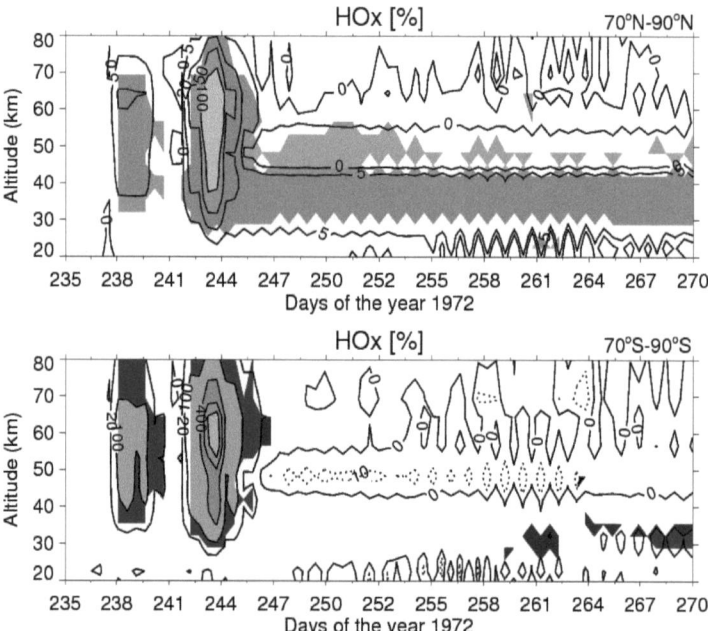

Figure 6.3: *Upper panel: Time-altitude profile of HO_x zonal mean change (in percent) for 70° - 90° N. Lower panel: same for southern hemisphere. Colors indicate areas with at least 95 % statistical significance. Contour levels for upper panel: -10, 0, 5, 20, 100, 500. Contour levels for lower panel: -10, 0, 20, 100, 400, 800, 1'200.*

Because the odd hydrogen species have not such a longevity like NO_x, the significant impact is quite short-term, only during the time when the particles are penetrating in the atmosphere.

Highly significant decreases of ozone are predicted for the upper and middle stratosphere with losses of about 80 % and 60 % in the southern and northern hemisphere, respectively. These losses are strongly correlated to the major increases in NO_x and HO_x visible in Figs. 6.2 and 6.3 causing the destruction of the ozone from 80 km reaching all the way down to 30 km. Below the depleted layer there is some indication of self-healing of ozone in the lower stratosphere.

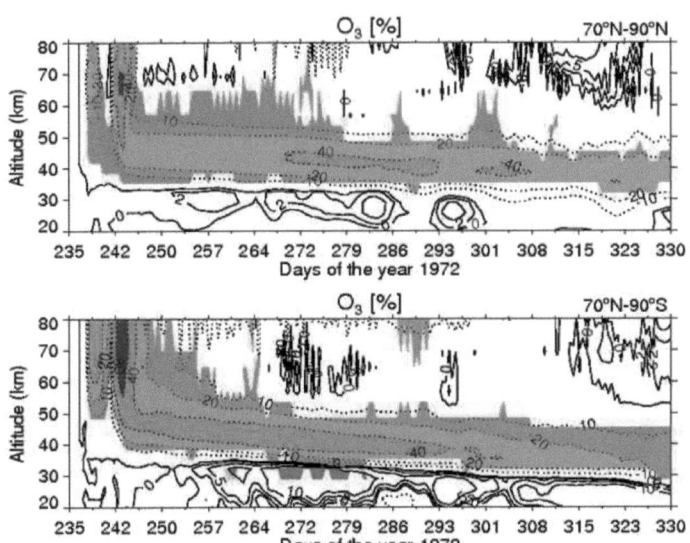

Figure 6.4: *Upper panel: Time-altitude profile of ozone zonal mean change (in percent) from 70° - 90° N. Lower panel: same for southern hemisphere. Colors indicate areas with at least 95 % statistical significance. Contour levels (%): -100, -80, -40, -20, -10, 0, 2, 5, 10, 50.*

There is a direct response in temperature following these ozone perturbations. The depletion in ozone will lead to a decrease in solar heating in the sunlit atmosphere which is visible in the upper panel in Fig. 6.5 for the NH shortly after the event has started. The model shows a significant cooling of more than 3 K at about 60 km. Conversely the SH, though showing a similarly strong cooling pattern, because of the lower insolation develops more fluctuations and therefore the first indications of significance appear only long after the impact.

Overall the results are similar in both hemispheres with an average decrease of about 3 K in the upper stratosphere/lower mesosphere. Albeit not significant, the lower and middle stratosphere shows a warming. In the lower stratosphere this is due to the higher levels of ozone, which lead to stronger heating. In the middle stratosphere (around 40 km) this is due to the higher levels of UV light penetrating from above. Therefore, as one could have expected, the division line between cooling above and warming below is not at the bottom of the layer depleted in ozone, but closer to its center.

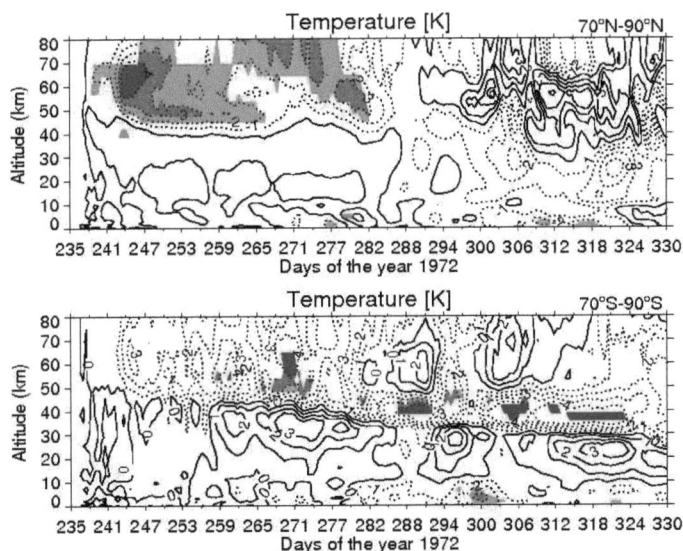

Figure 6.5: *Upper panel: Time-altitude profile of zonal mean temperature change (in Kelvin) for 70° - 90° N. Lower panel: same for the southern hemisphere. Colors indicate areas with at least 95 % statistical significance. Contour levels (K): -10, -4, -3, -2, -1, 0, 1, 2, 3, 5, 10.*

Figure 6.6 shows the changes averaged from August to November for the zonal wind for the northern and the southern hemisphere at 25 hPa (upper panel) and at 63 hPa (lower panel). Both altitude levels show in the NH an increase of up to 3 m/s in the region where the polar vortex is situated. The acceleration of the zonal wind is explained by the cooling of the polar upper stratosphere due to the polar ozone depletion induced by the SPE opposed to the warming in the middle stratosphere (see Fig. 6.5). These changes increase the temperature gradient, leading to the acceleration in agreement with the thermal wind balance (Limpasuvan et al., 2005). The SH shows a similar pattern to the NH but the increase in speed goes up to 5 m/s. The reason for this behavior can be again explained similar to the acceleration in the northern hemisphere.

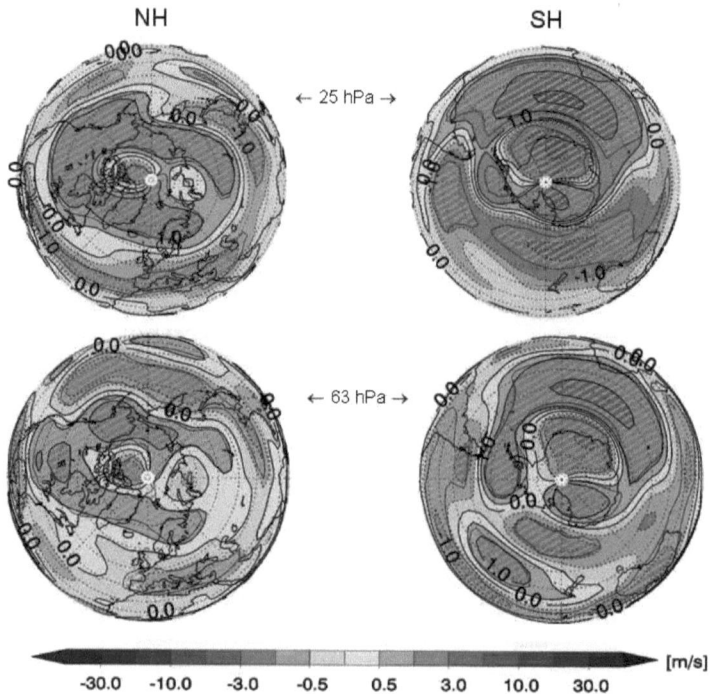

Figure 6.6: *Upper part: Polar stereographic projection of zonal wind changes at 25 hPa averaged from August to November (in m/s). Lower part: Zonal wind changes at 63 hPa averaged from August to November (in m/s). Left: northern hemisphere. Right: southern hemisphere. Hatched areas show 95 % statistical significance.*

The effects of the Carrington like event on total ozone are displayed in Fig. 6.7. Total ozone is reduced by a maximum of about 20 DU in October in the northern hemisphere. The SH shows a maximum reduction of about 40 DU in November. These maximum total ozone changes are not predicted to occur during the time when the SPE is happening; the transport of the enhanced NOx to lower altitudes and therefore higher ozone amounts allows more substantial total ozone impact. The collar like pattern from September to November in the northern hemisphere and from September to October in the southern hemisphere is explained that the polar vortex acts like a border for the transport of the ozone rich air from the midlatitudes. In November when the polar vortex in the SH breaks up the ozone rich air from the lower latitudes are able to enter the polar region which is seen in the lowermost picture on the right side.

SPE results

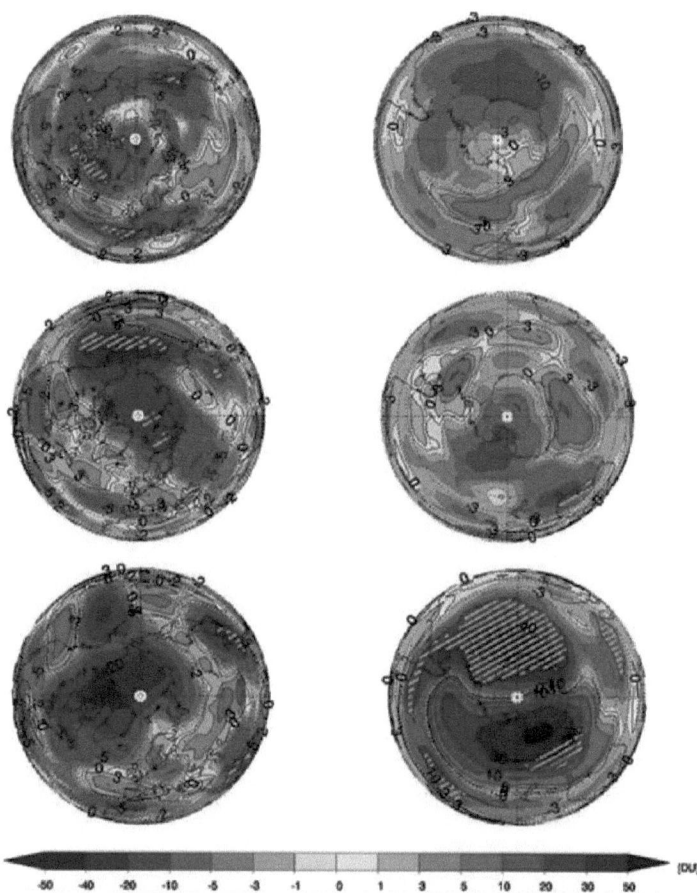

Figure 6.7: *Left row: Polar stereographic projection of changes in total ozone (in DU) for the northern hemisphere from September to November resulting from the Carrington Event. Right row: Same for the southern hemisphere. Hatched areas show 95 % statistical significance.*

Discussion and Conclusion

In the preceding sections we presented the influence of the August 1972 SPE, which is similar to the Carrington Event from 1859, on the atmospheric chemistry, temperature and dynamics. Based on the ^{10}Be isotope measurements in the ice core, the total fluence of energies >30 MeV was scaled to match the value given by Smart et al. (2006). The IR were calculated as described by Verronen et al. (2005) up to 0.01 hPa.

From the results presented in section 3 we draw the following conclusions. The influence of a Carrington like event on the chemistry and dynamics consists of NO_x and HO_x enhancement, depletion of ozone mixing ratios and total ozone column, cooling in the mesosphere and an acceleration of the zonal wind. Statistically significant effects down to the troposphere are negligible in the chemical and meteorological parameters (Figs. 6.2-6.6). The middle atmospheric changes due to the SPE are most pronounced shortly after the event happened.

The modeled impact on NO_x, HO_x and ozone is in reasonable agreement with the results of Rodger et al. (2008) and Thomas et al. (2007). Our results show a similar impact pattern compared to Rodger et al., (2008), i.e. the SH shows a larger response to the SPE than the NH. A difference concerning the longevity of the impact is visible, this is caused by the fact that we use a 3-D model with transport and dynamics implemented. Regarding the longevity of the impact, Thomas et al., (2007) show for ozone and the nitrogen species, that after 2 months, the depletion is still visible which is also shown in our plots.

Our results for temperature and dynamics are in good qualitative agreement with the paper of Jackman et al. (2007). They show that shortly after the event happened, the southern hemispheric polar region has a decrease in temperature from upper down to lower mesosphere, similar to our results for the northern hemispheric polar region. The difference among their results and ours is in the intensity of the changes. For the temperature a decrease of more than 3 K is shown in this work while Jackman et al. (2007) depict a decrease of up to 2 K. The fact that our results show a larger effect can be due to the intensity of the solar proton event. The Carrington like event which is presented in this paper represents an event that is more intense than the SPE of Oct/Nov 2003.

The qualitative agreement of our results, modeled with the 3-D CCM SOCOL, with the changes in NO_x, HO_x, ozone, temperature and dynamics, obtained by Rodger et al., (2008), Thomas et al. (2007) and Jackman et

al. (2007), allows us to conclude, that we can strengthen the statement that this solar proton event has had intense interaction in a broad altitude range starting from 80 km going down to almost 30 km.
From these conclusions we can say that such an extreme SPE can heavily influence the atmospheric chemistry, temperature and dynamics (see Fig(s). 6.2 to 6.7). Therefore it is important to analyze the impact of energetic particles with a 3-D CCM to be sure that the dynamical aspects and transport are taken into account.

References

Crutzen, P.J.: The influence of nitrogen oxides on the atmospheric ozone content, Quart. J. R. Met. Soc., 96, 320 325, 1970.

Egorova, T., Rozanov, E., Zubov, V. and Karol, I.L.: Model for Investigating Ozone Trends (MEZON), Izvestiya, Atmos. Ocean. Phys., 39, 277-292, 2003.

Egorova, T., Rozanov, E., Zubov, V., Manzini, E., Schmutz, W. and Peter, T.: Chemistry-climate model SOCOL: a validation of the present-day climatology, Atmos. Chem. Phys., 5, 1557-1576, 2005.

Eyring, V., et al.: Assessment of temperature, trace species, and ozone in chemistry-climate model simulations of the recent past, J. Geophys. Res., 111, D22308, doi:10.1029/2006JD007327, 2006.

Eyring, V., et al.: Multimodel projections of stratospheric ozone in the 21st century, J. Geophys. Res., 112, D16303, doi:10.1029/2006JD008332, 2007.

Herman, A.J. and Goldberg, A.R.: Initiation of non-tropical thunderstorms by solar activity, J. Atmos. Terr. Phys., 40, 121 134, 1978.

Jackman, C.H., Roble G.R. and Fleming, E.L.: Mesospheric dynamical changes induced by the solar proton events in October-November 2003, Geophys. Res. Lett., 34, L04812, doi:10.1029/2006GL028328, 2007.

Jackman, C.H., Douglass, A.R., Rood, R.B. and McPeters, R.D .: Effect of Solar Proton Events on the middle atmosphere during the past two solar cycles as computed using a two-dimensional Model, J. Geophys. Res., 95, D6, 7417 7428, 1980.

Krivolutsky, A., Bazilevskaya, G., Vyushkova, T. and Knyazeva, G.: Influence of cosmic rays on chemical composition of the atmosphere: data analysis and photochemical modeling, Phys. and Chem. of the Earth, 27, 471 476, 2002.

Limpasuvan, V., Hartmann, D.L., Thompson, D.W.J., Jeev, K. and Yung, Y.L.: Stratosphere-troposphere evolution during polar vortex intensification, J. Geophys. Res., 110, D24101, doi:10.1029/2005JD006302, 2005.

Manzini, E., McFarlane, N.A., McLandress, C.: Impact of the Doppler spread parameterization on the simulation of the middle atmosphere circulation using the MA/ECHAM4 general circulation model, J. Geophys. Res. Atmos., 102, D22, 25751 25762, 1997.

Nicolet, M.: On the production of nitric oxide by cosmic rays in the mesosphere and stratosphere, Planet. Space Sci., 23, 637 649, 1975.

Porter, H.S., Jackman, C.H., Green, A.E.S.: Efficiencies for production of atomic nitrogen and oxygen by relativistic proton impact in air, J. Chem. Phys., 65, No.1, 1976.

Prather, M.J.: Numerical Advection by conservation of 2nd-order moments, J. Geophys. Res. Atmos., 91, D6, 6671 6681, 1986.

Rodger, C.J., Verronen, P.T., Clilverd, M.A., Seppl, A. and Turunen, E.: Atmospheric impact of the Carrington event solar protons, J. Geophys. Res., 113, D23302, doi:10.1029/2008JD010702, 2008.

Rozanov, E., Schlesinger, M.E., Zubov, V., Yang, F. and Andronova, N.G.: The UIUC three-dimensional stratospheric chemical transport model: Description and evaluation of the simulated source gases and ozone, J. Geophys. Res., 104, 11,755-11,781, 1999

Schraner, M., Rozanov, E., Schnadt-Poberaj, C., Kenzelmann, P., Fischer, A., Zubov, V., Luo, B.P., Hoyle, C., Egorova, T., Fueglistaler, S., Brnnimann, S., Schmutz W. and Peter, T.: Chemistry climate model SOCOL: version 2.0 with improved transport and chemistry/ microphysics schemes, Atmos. Chem. Phys., 8, 19, 5957-5974, 2008.

Simpson, J.A.: Elemental and Isotopic composition of the Galactic Cosmic Rays, Ann. Rev. Nucl. Part. Sci. 33, 323-381, 1983.

Smart, D.F., Shea, M.A. and McCracken, K.G.: The Carrington event: Possible solar proton intensity-time profile, Adv. Space Res., 38, 215-225, 2006.

Solomon, S., Rusch, D.W., Gerard, J.-C., Reidt, G.C. and Crutzen, P.J.: The effect of particle precipitation events on the neutral and ion chemistry of the middle atmosphere: II. Odd Hydrogen, Planet. Space Sci., 29, No. 8, 885 892, 1981.

Swider, W. and Keneshea, T.J.: Decrease of ozone and atomic oxygen in lower mesosphere during a PCA event, Planet. Space Sci., 21, No.11, 1969 1973, 1973.

Thomas, B.C., Jackman, C.H. and Melott, A.L.: Modeling atmospheric effects of the September 1859 solar flare, Geophys. Res. Lett., 34, L06810, doi:10.1029/2006GL029174, 2007.

Usoskin, I.G. and Kovaltsov, A.: Cosmic ray induced ionization in the atmosphere: Full modeling and practical applications, J. Geophys. Res., 111, D21206, doi:10.1029/2006JD007150, 2006.

Verronen, P.T., Seppl, A., Clilverd, M.A., Rodger, C.J., Kyrl, E., Enell, C.-F., Ulich, Th. and Turunen, E.: Diurnal variation of ozone depletion during the October-November 2003 solar proton events, J. Geophys. Res., 110, A09S32, doi:10.1029/2004JA010932, 2005.

Vitt, F.M. and Jackman, C.H. :A comparison of sources of odd nitrogen production from 1974 through 1993 in the Earths middle atmosphere as calculated using a two-dimensional model, J. Geophys. Res., 101, D3, 6729 6739, 1986.

Williamson, D.L. and Rasch, P.J.: Two-Dimensional Semi-Lagrangian Transport with shape-preserving interpolation, Mon. Weather Rev., 117, 1, 102 129, 1989.

Zubov, V.A., Rozanov, E.V., Schlesinger, M.E.: Hybrid scheme for three-dimensional advective transport, Mon. Weather Rev., 127, 6, 1335 1346, 1999.

Chapter 7

GCR, SPE and LEE simultaneously

7.1 Introduction

Galactic cosmic rays, solar protons and energetic electrons are energetic particles which originate from outside of the solar system, from the Sun and from the Van Allen belts (Bazilevskaya et al., 2008; Jackman et al., 2008; Hudson et al., 2008). The flux of these energetic particles to the Earth varies as a result of the solar activity, i.e. it is modulated by the solar wind and the magnetic field. When the energetic particles enter the Earth's atmosphere they collide with the ambient atmospheric gas molecules, thereby ionizing them. In this process they may produce secondary particles which can be energetic enough to contribute themselves to further ionization of the neutral gases. This leads to the development of an ionization cascade or shower. The intensity and penetration depth of the cascade depends on the energy of the primary particle. Cascades of particles with several hundred MeV of kinetic energy may reach the ground (Bazilevskaya et al., 2008).

Due to their charge, the energetic particles are additionally deflected by the geomagnetic field. Almost all particles can penetrate into the polar region, whereas only the energetic particles with energies above 15 GeV are able to penetrate the atmosphere near the equator.

First models of the cosmic ray induced ionization (CRII) were (semi)empirical (e.g., OBrien, 1970; Heaps, 1978) or simplified analytical (Vitt Jackmann, 1996, OBrien, 2005). State-of-the-art models (Usoskin et al., 2004; Desorgher et al., 2005; Usoskin Kovaltsov, 2006) are based on Monte-Carlo simulations of the atmospheric cascade and can provide 3-D time dependent computations of the CRII.

The ionization rates for the solar protons used in models have usually been computed out of observations, i.e. satellites or ground based measurements (e.g. McPeters et al., 1981; Jackman et al., 1995; Seppälä et al., 2008). Recently, Baumgärtner et al. (2010) proposed a parameterization for the ionization rates for the solar protons. This parameterization is based on flux measurements by instruments on the IMP and GOES satellites.

The ionization due to the energetic electrons have been usually calculated with the flux taken from observations (Callis et al., 1998; Rozanov et al., 2005; Clilverd et al., 2009). Baumgärtner et al. (2009) gives a parameterization for the LEE based on measurements of the geomagnetic activity.

The ionization induced by the energetic particle precipitation leads to the production of odd nitrogen (Jackman et al., 2008). For example, fast secondary electrons (e*) can dissociate the nitrogen molecule and almost all of the N atoms in the ^{2}D state react with O_2, producing nitric oxide (7.1 and 7.2).

$$N_2 + e^* \rightarrow 2N(^2D) + e \qquad (7.1)$$
$$N(^2D) + O_2 \rightarrow NO + O \qquad (7.2)$$

The Galactic cosmic rays are estimated to lead to the production of 3.0 to 3.7 $\cdot 10^{33}$ molecules of odd nitrogen per year in the global stratosphere, which amounts to about 10 % of the NO_x production following N_2O oxidation. However, the solar protons can produce an amount ranging from 9.0 $\cdot 10^{30}$ to 3.3 $\cdot 10^{33}$ molecules of odd nitrogen per year globally in the stratosphere (Vitt Jackman, 1996). In comparison, the northern polar/subpolar region ($> 50^o$) is believed to be supplied with NO_x in equal amounts by the energetic particle precipitation and by N_2O oxidation. In the deep polar winter stratosphere, when air masses experience sunlit periods only infrequently and photolysis of HNO_3 becomes negligible, the precipitation of the energetic particles becomes the only source of NO_x, revealing the importance in high latitudes.

Below the mesopause, where water cluster ions can be formed, the ionization induced by the GCRs, SPEs and EEPs contribute to the formation of HO_x radicals. For example, molecular oxygen ions (O_2 $^+$) can via attachment of molecular oxygen form O_4 $^+$, which reacts with water (7.3). Equation 7.4 shows that this hydrated ion quickly hydrates further producing OH (Egorova et al., 2010).

$$O_4^+ + H_2O \rightarrow O_2^+H_2O + O_2 \qquad (7.3)$$
$$O_2^+H_2O + H_2O \rightarrow H_3O^+OH + O_2 \rightarrow H_3O^+ + OH + O_2 \qquad (7.4)$$

7.1 Introduction

HO_x production via the energetic particles competes with the most important source for HO_x in the atmosphere, which is the photolytically driven oxidation of water vapor (H_2O) by excited oxygen atoms, $O(^1D)$, which are themselves produced from ozone photolysis.

However, during polar night, HO_x is mainly produced by the GCRs, SPEs and EEPs given that only little UV radiation is available for $O(^1D)$ production.

Here we study the effect of the ionization due to the GCRs, SPEs and EEPs using the global chemistry-climate model (CCM) SOCOL focusing on the estimation of the sensitivity of the model chemistry, temperature and dynamics on the influence of NO_x and HO_x from the surface up to 0.01 hPa barometric pressure (altitude of approximately 80 km).

The influence of the energetic particles all together to the atmospheric chemistry and dynamics using a 3-D CCM is a novelty. The author has not found any published paper dealing with the same topic.

The description of the CCM SOCOL and the energetic particles induced modeling are described in Section 7.2. The results are presented in Section 7.3.

7.2 Description of the Model and experimental setup

7.2.1 Chemistry-Climate Modeling

The description of the CCM SOCOL used in this experiment can be looked up in chapter 4 of this thesis.

7.2.2 Energetic Particles Induced Modeling

The ionization due to the Galactic cosmic rays has been parameterized using the recently developed CRAC:CRII (Cosmic Ray induced Cascade: Application for Cosmic Ray Induced Ionization) model (Usoskin et al., 2004; Usoskin Kovaltsov, 2006) extended toward the upper atmosphere (Usoskin, Kovaltsov Mironova, 2010). The model is based on a Monte-Carlo simulation of the atmospheric cascade and reproduces the observed data within 10 % accuracy in the troposphere and lower stratosphere (Bazilevskaya et al., 2008; Usoskin et al., 2009). In the mesosphere the agreement between observed and simulated ionizations rates are not good because the ionization by other sources (solar radiation, precipitating soft particles of magnetospheric origin, etc.) becomes at least as important as from GCRs. The results of the CRAC:CRII model are parameterized (see Usoskin et al., 2005) to give ion pair production rate as a function of the altitude (quantified via the barometric pressure), geomagnetic latitude (quantified via geomagnetic cutoff rigidity) and solar activity (quantified via the modulation potential θ).

For the solar protons, daily averaged ionization rates from 1963 to 2008, valid from 60° to 90° as functions of pressure between 888 hPa (1 km) and $8 \cdot 10^{-5}$ hPa (115 km) from the SOLARIS (Solar Influence for SPARC) website www.geo.fu-berlin.de/en/met/ag/strat/forschung/SOLARIS/Input_data/index.html have been taken and introduced to our model.

The low energetic electrons have been paramterized using Baumgärtner et al., (2009). Because the precipitation of the electrons depend on the magnetic activity of the Sun, Baumgärtner et al., (2009) used the Ap index to deal with this problem. It has been shown by Funke et al., (2005) that the impact of the LEE are confined to the vortex. Therefore a minimum absolute latitude of 55° has been used, i.e. that the energetic electrons do not enter in the atmosphere lower than this latitude. The fact that the low energetic elec-

7.2 Description of the Model and experimental setup

trons lose their energy already in the thermosphere, sponge layers had to be created in our model where the initial ionization could take place. Below the sponge layers, the NO_x species are mostly transported downward. To differ between the northern hemisphere (NH) and the southern hemisphere (SH), Baumgärtner et al., (2009) used the solstice in their parameterization, i.e. depending on which hemisphere gets ionized, the day of the solstice changes. The ionization rates of the above mentioned particles cannot be directly used in CCM SOCOL which has no explicit treatment of ion chemistry, therefore it is necessary to convert the ionization into the NO_x and HO_x production. Following Porter et al. (1976), 1.25 NO_x molecules are produced per ion pair, and 45 % of this NO_x production is assumed to yield ground state atomic nitrogen, while 55 % is assumed to go into $N(^2D)$ with instantaneous conversion to NO (see Introduction). The production of HO_x has been studied by Solomon and Crutzen (1981) with a 1-D time-dependent model of neutral and ion chemistry. They parameterized the number of odd hydrogen particles produced per ion pair as a function of altitude and ionization for daytime, polar summer conditions of temperature, air density and solar zenith angle. We implement these parameterizations in the CCM SOCOL to take into account the production of NO_x and HO_x induced through the GCRs, SPEs and EEPs from the ground up to the height of 0.01 hPa barometric pressure (altitude of 80 km).

For this study, we have carried out two 46-year long runs of CCM SOCOL v2.0 from 1960 to 2005. The control run has been performed without the influence of the energetic particles, while the experiment run includes GCRs, SPEs and LEEs using the IR given by Usoskin et al., (2010), the SOLARIS website and Baumgärtner et al., (2009) up to 0.01 hPa. The first two years of the runs have been neglected for the analysis to avoid possible spin up problems of the model.

7.3 Results

Figures 7.1 - 7.7 show annual mean response of the zonal mean NO_x, HO_x, HNO_3 and ozone to the GCRs, SPEs and EEPs. Figure 7.8 shows monthly mean zonal mean response for ozone and zonal wind for November. The annual mean zonal mean response of total ozone is shown in Fig. 7.9

The Galactic cosmic rays, solar protons and energetic electrons produce substantial amount of NO_x during all seasons (not shown). The annual mean shows that the simulated NO_x increase is most intense in the upper to lower mesosphere (see Fig. 7.2). The increase exceeds 30 ppb or 1500 % for a region extending from the south pole to north pole. In the stratosphere, the maximum increase of NO_x goes up 3.4 ppb or 40 % covering the region from $40°$ N to $90°$ N and $50°$ S to $90°$ S. The increase in the troposphere exceeds 20 % or 10 ppt for a region extending from the south pole to the north pole at a height of about 10 km.

The difference between southern to northern hemisphere is explained by the fact that more NO_x is produced anthropogenically in the NH than in the SH, therefore the relative difference in the northern hemisphere is smaller and less significant than in the remote regions in the SH.

The reason why the NO_x increase is less pronounced closer to the surface can be explained that only particles with energies up to hundreds of MeV are able to penetrate to this altitude. In the upper part of the mesosphere the GCRs, SPEs and EEPs are all able to ionize the air which is clearly visible by the increase of more than 30 ppb (see Fig. 7.2). Close to the surface, only the Galactic cosmic rays are capable to produce odd nitrogen because these particles are energetic enough to penetrate down to this level.

7.3 Results

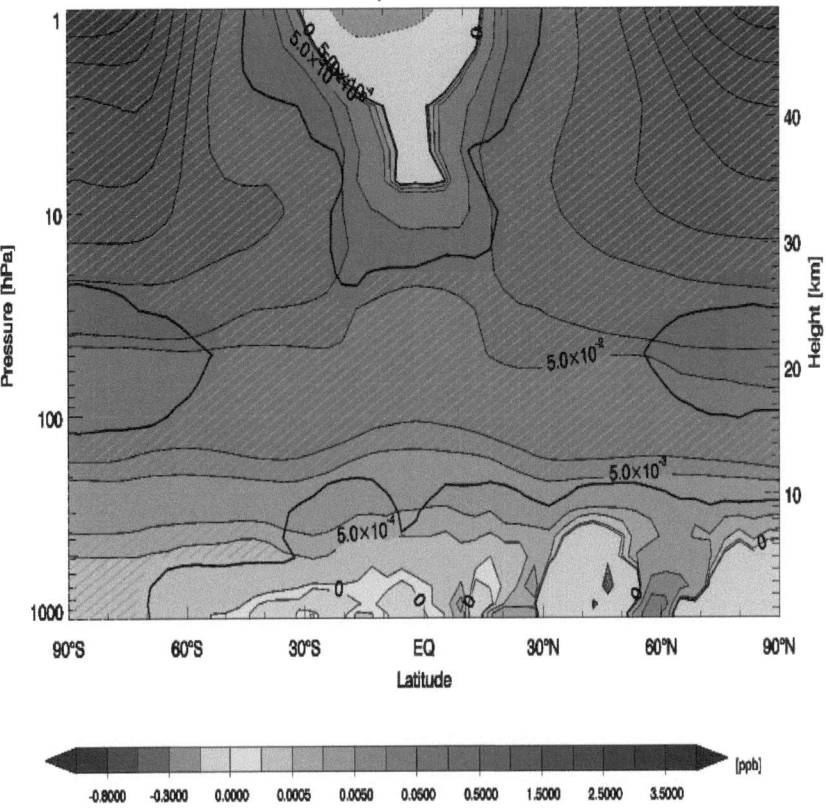

Figure 7.1: *Annual mean changes of zonal mean NO_x, ($[NO_x]$exp-$[NO_x]$control), in ppb ($[NO_x] = [NO] + [NO_2]$). Results are averaged from 1962-2005. Solid contours indicate positive, dotted contours negative changes. Hatched areas (enclosed by solid contours) indicate changes with at least 95 % statistical significance*

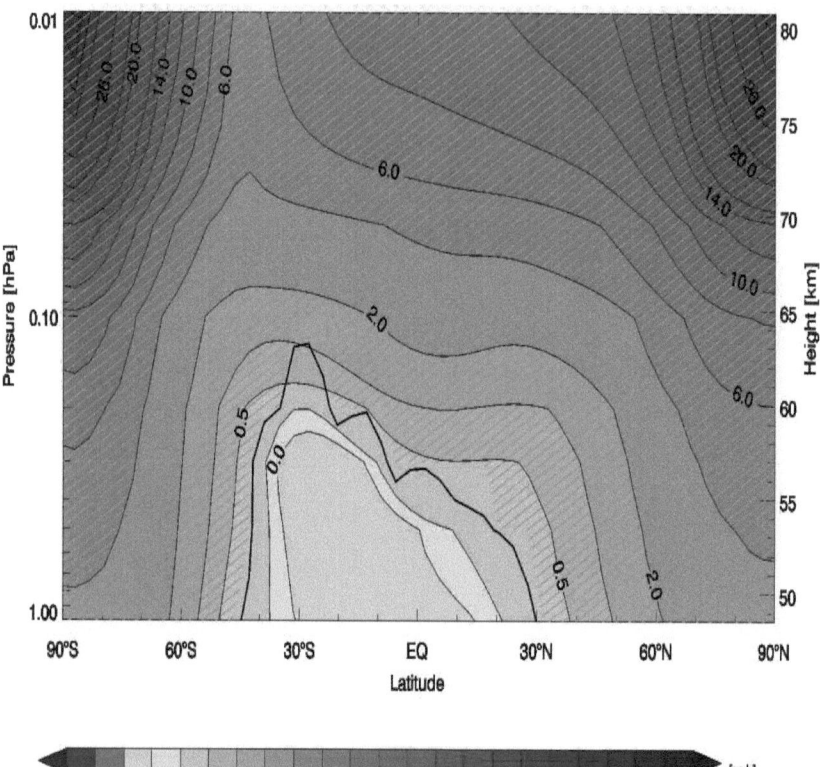

Figure 7.2: *Annual mean changes of zonal mean NO_x, ($[NO_x]exp-[NO_x]control$), in ppb ($[NO_x] = [NO] + [NO_2]$). Results are averaged from 1962-2005. Solid contours indicate positive, dotted contours negative changes. Hatched areas (enclosed by solid contours) indicate changes with at least 95 % statistical significance*

7.3 Results

Figure 7.3 represents the response of annual mean zonal mean HO_x to GCRs, SPEs and EEPs. The HO_x production does not result in a statistically significant change throughout the whole atmosphere except a broad area of significant odd hydrogen reduction in the tropical/mid-latitude UTLS. The hatched area shows a maximum decrease of about 6 % or 0.05 ppt over the northern hemispheric mid-latitudes at an altitude of about 20 km. This broad area of HO_x decrease coincides with a region of high NO_x enhancements and can be explained by the more intensive removal of OH via $OH + NO_2 + M \rightarrow HNO_3 + M$, resulting in a significant HNO_3 increase of about 12 % or up to 0.4 ppb (Fig. 7.5).

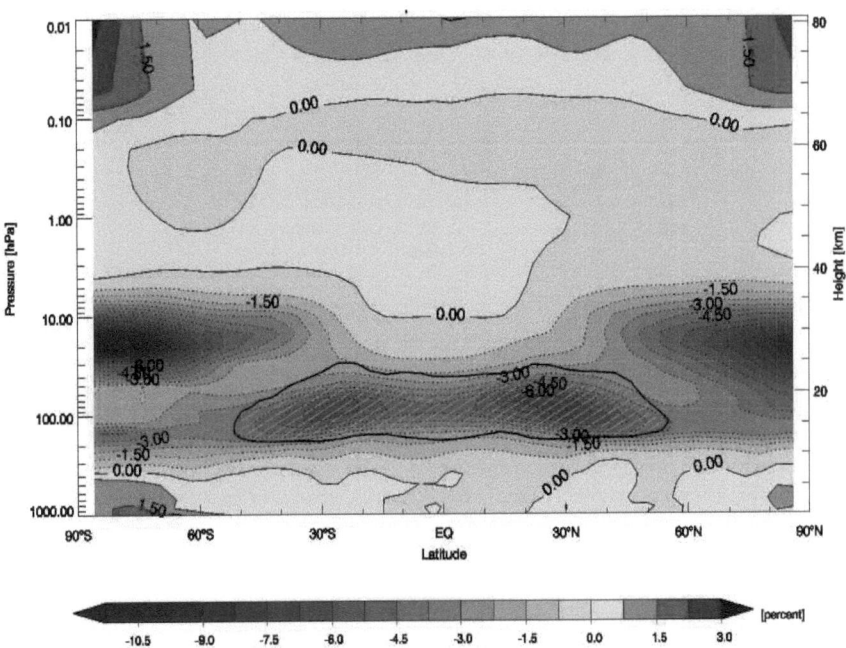

Figure 7.3: *Annual mean changes of zonal mean HO_x, $([HO_x]exp-[HO_x]control)/[HO_x]control$, in percent ($[HO_x] = [H] + [OH] + [HO_2]$). Results are averaged from 1962-2005. Hatched areas (enclosed by solid contours) indicate statistically significant changes with at least 95 %.*

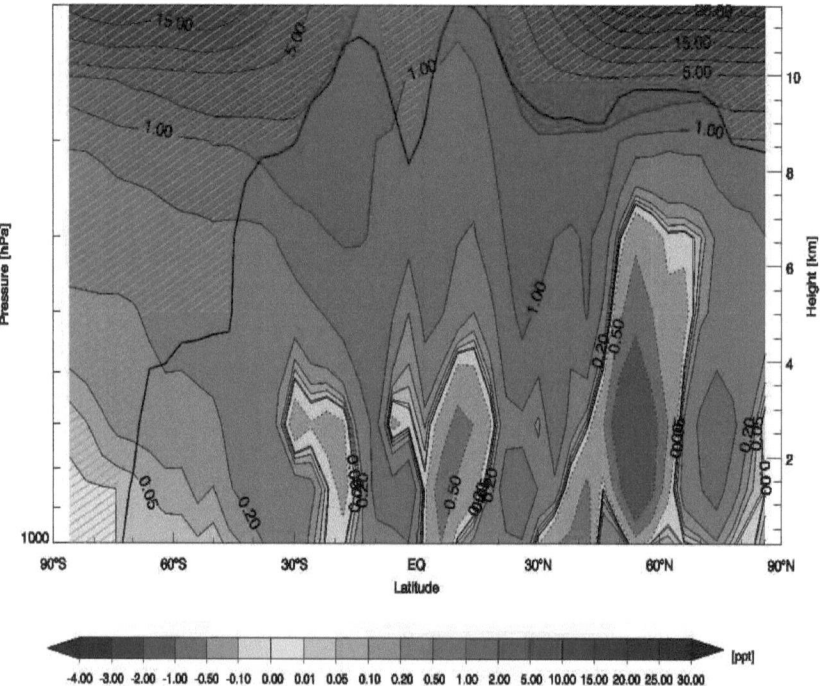

Figure 7.4: *Annual mean changes of zonal mean HNO_3, ($[HNO_3]exp$-$[HNO_3]control$), in ppt from 1000 hPa up to 200 hPa. Results are averaged from 1962-2005. Solid contours indicate positive, dotted contours negative changes. Hatched areas (enclosed by solid contours) indicate changes with at least 95 % statistical significance*

Figures 7.4 - 7.6 show the changes for HNO_3 due to the GCRs, SPEs and LEE together.

Figure 7.4 reveals that the HNO_3 significantly increases from surface up to tropopause with a maximum of approximately 12 ppt.

The northern hemispheric upper troposphere shows a significant increase of up to 20 ppt. The reason why the lower levels in the NH show no significant increase is explained when looking at Fig. 7.1.

7.3 Results

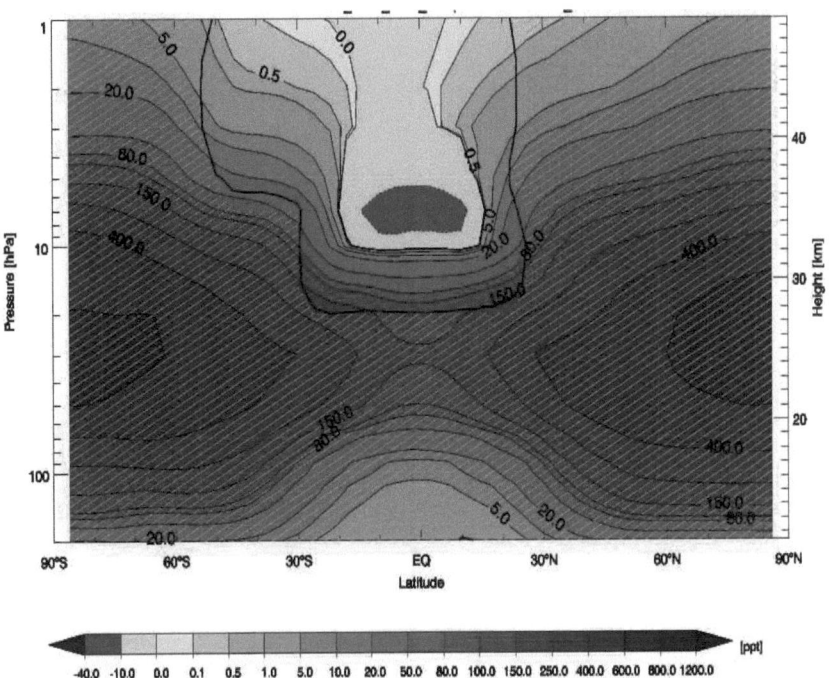

Figure 7.5: *Annual mean changes of zonal mean HNO_3, ($[HNO_3]exp-[HNO_3]control$), in ppt from 200 hPa up to 1 hPa. Results are averaged from 1962-2005. Solid contours indicate positive, dotted contours negative changes. Hatched areas (enclosed by solid contours) indicate changes with at least 95 % statistical significance*

The additional input of odd nitrogen through the GCRs in the SH polar region shows a significant impact caused by the fact that this area is NO_x-limited.

The most intense changes due to the energetic particles is shown in Fig. 7.5. Both hemispheres show an increase of about 600 ppt in an altitude range from 18 km to 38 km.

Figure 7.6 which shows the mesosphere going from 1 hPa up to 0.01 hPa revealing that in this altitude range the changes for HNO_3 due to the en-

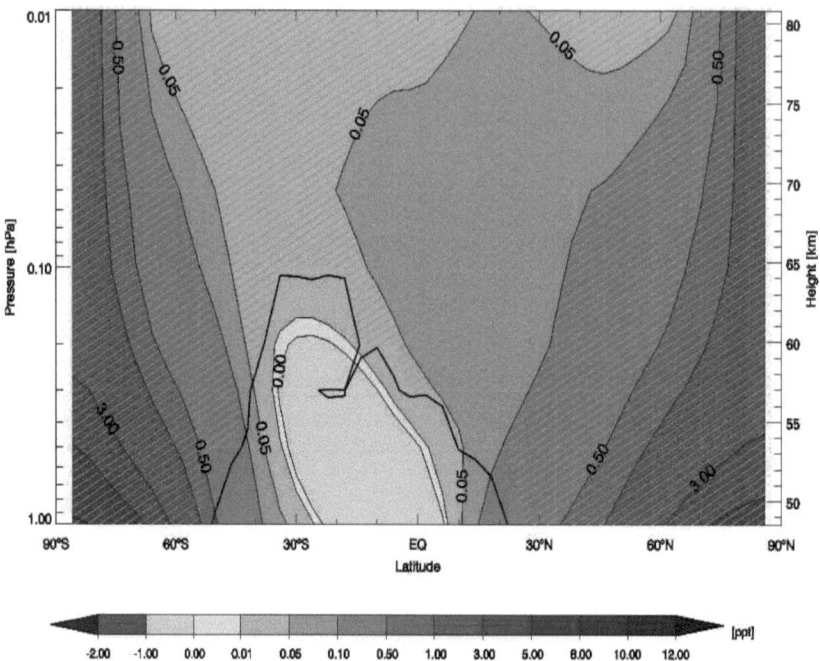

Figure 7.6: *Annual mean changes of zonal mean HNO_3, ([HNO_3]exp-[HNO_3]control), in ppt from 1 hPa up to 0.01 hPa. Results are averaged from 1962-2005. Solid contours indicate positive, dotted contours negative changes. Hatched areas (enclosed by solid contours) indicate changes with at least 95 % statistical significance*

ergetic particles are weakest. The maximum increase of 8 ppt is visible in the southern hemisphere close to the stratopause. The northern hemisphere shows a maximum increase of about 5 ppt.

Significant increase of NO_x in the southern hemispheric troposphere leads to the statistically significant ozone enhancement. As mentioned in chapter 3.2, ozone photochemistry in the SH is in large parts NO_x-limited, so that the ionization through the GCRs relaxes this limitation leading up to 2.5 % or 3 ppb ozone increase. Another significant increase is visible at about 20

7.3 Results

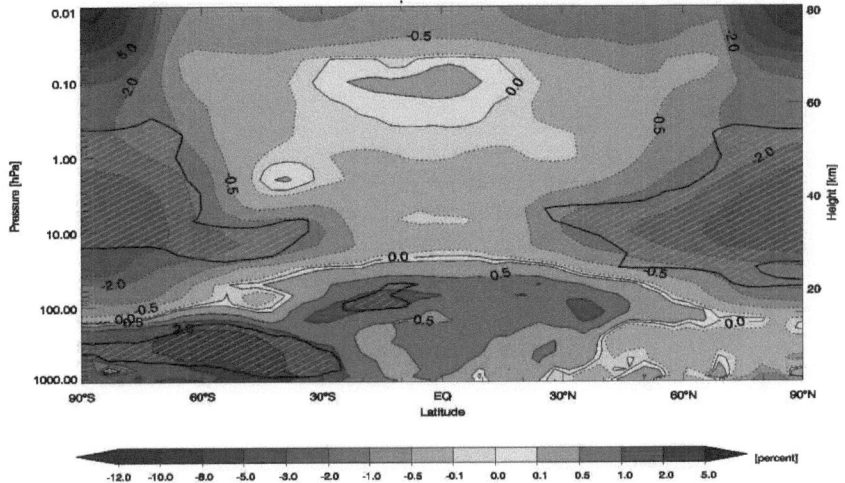

Figure 7.7: *Annual mean changes of zonal mean ozone, ([O_3]exp-[O_3]control)/[O_3]control, given in percent. Results are averaged from 1962-2005. Solid contours indicate positive, dotted contours negative changes. Hatched areas (enclosed by solid contours) indicate changes with at least 95 % statistical significance.*

km from $10°$ S to $25°$ S of 1 % or 10 ppb (see Fig 7.7). The reason why the relative increase of 1 % at 20 km over the equatorial region is higher than the 2.5 % change in the southern hemispheric midlatitudes at about 5 km is that the concentration of ozone is higher in the stratosphere than in the troposphere.

Conversely, in the northern and southern polar stratosphere a significant ozone decrease of up to 3 % or up to 0.2 ppm is depicted. This decrease is typical for the direct impact of the solar protons which can produce NO_x directly in this altitude range. The energetic electrons also participate to the destruction of O_3. The electrons are not destroying directly the ozone because the deposition happens in the upper mesosphere. The NO_x produced through the EEPs has been transported down to the stratosphere where it can contribute to the O_3 destruction together with the NO_x produced through the solar protons.

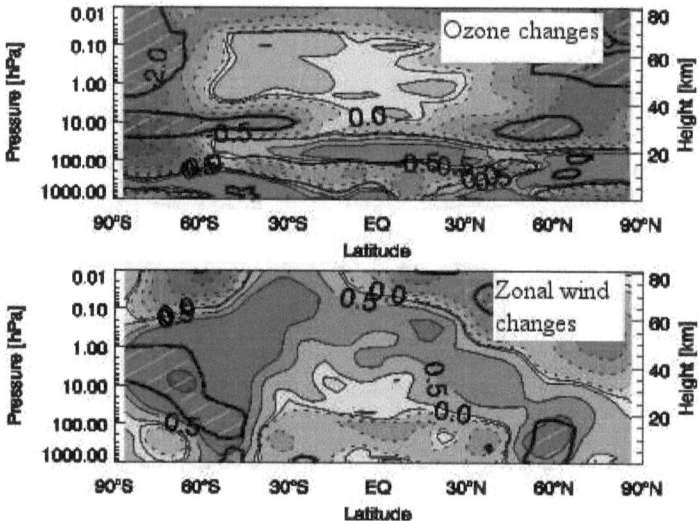

Figure 7.8: *Upper panel: Monthly mean changes for zonal mean ozone for November given in percent. Lower panel: Monthly mean changes of zonal mean zonal wind for November given in m/s. Hatched areas (enclosed by solid contours) indicate changes with at least 95 % statistical significance.*

The panels in Fig. 7.8 show the influence of the energetic particles on the monthly mean zonal wind (lower panel) and ozone (upper panel) for November. A significant increase of up to 5 m/s in the SH polar region from about 1 hPa down to 100 hPa is visible. The acceleration is caused by the cooling of the polar stratosphere due to ozone depletion (Fig 7.8, upper panel). This cooling increases the temperature gradient between the equatorial and the polar regions leading to acceleration of the zonal wind in agreement with the thermal wind balance.

7.3 Results

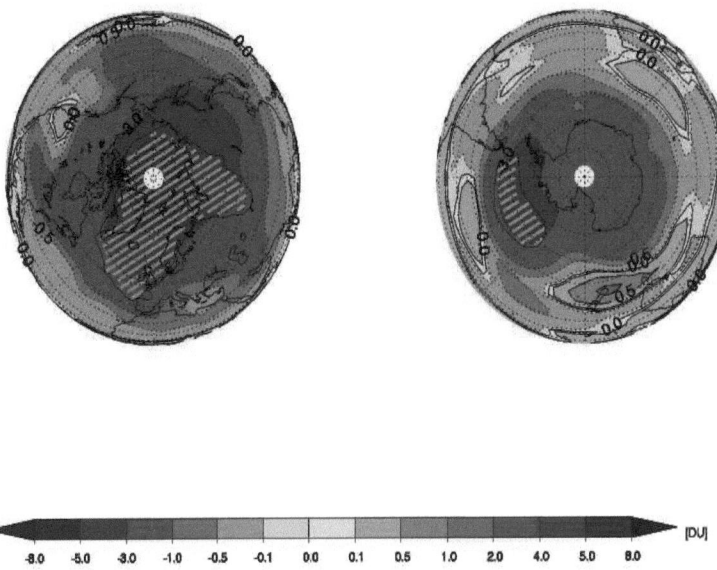

Figure 7.9: *Annual mean changes for total ozone given in DU due to the GCRs, SPEs and LEEs. Results are averaged from 1962 to 2005. Left panel represents the NH whereas the SH is on the right side. Hatched areas (enclosed by solid contours) indicate changes with at least 95 % statistical significance.*

The total ozone changes due to the GCRs, SPEs and LEEs are presented in Fig. 7.9. Total ozone is reduced by a maximum of about 5 DU in the northern hemisphere annually averaged from 1962 to 2005 (left panel). The SH shows an annually averaged maximum reduction of about 3 DU during the same timespan (right panel). The maximum total ozone changes are not predicted to occur during the time when the energetic particles are entering the Earth's atmosphere; the transport of the enhanced NO_x to lower altitudes and therefore higher ozone amounts allows more substantial total ozone impact.

7.4 Conclusion

Our calculations indicate that the NO_x increase due to the ionization of the energetic particles is most intense from 80 km down to 60 km (see Fig. 7.2). The increase exceeds 30 ppb or 1500 % for a region extending from the south pole to north pole. In the stratosphere, the maximum increase of NO_x goes up 3.4 ppb or 40 % covering the region from $40°$ N to $90°$ N and $50°$ S to $90°$ S. The growth of odd nitrogen in the troposphere exceeds 20 % or 10 ppt for a region extending from the south pole to the north pole at a height of about 10 km.

HO_x shows at the same level of significance a maximum decrease of about 6 % or 0.05 ppt over the northern hemispheric mid-latitudes at an altitude of about 20 km.

The maximum changes for HNO_3 of about 0.5 ppb, caused by the GCRs, SPEs and EEPs are visible from the south pole to roughly $30°$ S between 100 hPa and 6 hPa. The northern hemisphere has a maximum increase that exceeds 0.7 ppb visible from the north pole to $30°$ N.

GCR related ozone increases from the South Pole to $30°$ S up to 2.5 % is depicted from nearly 900 hPa to 300 hPa. A significant increase in ozone is visible at about 20 km from $10°$ S to $25°$ S about of 1 % or 10 ppb. An ozone decrease induced through the solar protons and energetic electrons of up to 3 % or up to 0.2 ppm is depicted in the northern and southern polar stratosphere starting at an altitude of 55 km going down to almost 20 km with a maximum decrease visible in a height range between 45 km and 30 km. The modification of total ozone of about 5 DU or 2 % in the northern hemisphere and 3 DU or 1 % in the southern hemisphere emphasizes that the additional NO_x produced through the interaction of the energetic particles with the neutral atmosphere is having an impact on the chemistry due to the longevity of the odd nitrogen.

The dynamical changes, represented by the monthly mean changes for the zonal wind in November, show an acceleration in the SH polar region of up to 5 m/s in an altitude range from 50 km down to 15 km. These changes induced by the decrease in ozone, reveal that the changes in the atmospheric chemistry can have further impact on the dynamics.

As a concluding remark it can be said that the impact due to the energetic particles on the above mentioned species are not negligible. Therefore, it is important having the GCRs, SPEs and EEPs parameterized in model runs that deal with the atmospheric chemistry and dynamics.

Chapter 8
Conclusion and Outlook

8.1 Conclusion

This thesis has focused on the influence of energetic particles (GCRs, SPEs and EEPs) on atmospheric chemistry and climate from the ground level up to the mesopause. To achieve this goal, several simulations with the CCM SOCOL v2.0 have been performed to investigate the relation between forced and natural variability of the climate system. The results from SOCOL are in very good agreement with other state-of-the-art models as well as with observational data.

The output has been validated against observational and modeled data to detect model biases. A direct comparison with the observational and modeled data revealed that the results obtained with SOCOL v2.0 are in good agreement regarding the chemical species and the dynamics. Special emphasis was set on the investigation of the influence of the additonal NO_x and HO_x caused by the energetic particles over the entire height range of SOCOL and the influence on ozone, temperature and dynamics.

Our findings show that the results vary greatly among the different types of particle precipitation. The Galactic cosmic rays, for example, which are the particles with the highest energies, have their significant impact on the atmospheric chemistry and dynamics mostly throughout the stratosphere and in the troposphere. The largest changes caused by the solar protons affect the stratosphere and the mesosphere in the polar regions. The effect in the troposphere is negligible because the protons are not energetic enough to penetrate down to this altitude. The particles with the lowest energies, the energetic electrons, lose most of their energy in an altitude range between 70 km and 100 km. Therefore, the direct changes induced by the electrons are mostly restricted to the mesosphere. However, the additional NO_x produced

Conclusion and Outlook

by the electrons can get transported down to lower altitudes where it can participate in further reactions. Because the lifetime of HO_x is less than the lifetime of NO_x, the downward transportation of the odd hydrogen species is negligible.

The different parameterizations give varying results for the same energetic particles, for example, the GCRs show different results for NO_x and ozone (see Chapter 5.1). This emphasizes the importance of using an appropriate parameterization going down to the Earth's surface.

Modeling the solar proton event from October/November 2003 with the IR taken from the SOLARIS webiste and from Wissing & Kallenrode, (2009) reveals that the results are in good agreement regarding NO_x, HO_x and ozone. The only difference is that the changes due to the ionization rates by Wissing & Kallenrode (2009) are more pronounced than the ones modeled with the data from the SOLARIS website. The reason for this is clearly visible when looking at Figs. 4.4 and 4.5. These figures show that the ionization rates calculated with AIMOS are more pronounced during this event.

For modeling the energetic electrons, two different approaches have been used in this thesis. One approach has used the parameterization of Baumgärtner et al., (2009) for LEEs, the other the parameterization of Rozanov et al., (2005) which includes LEE as well as HEE. Both parameterizations have advantages and deficiencies. The IR calculated with Baumgärtner et al. (2009) can be used in simulations that last for several years because the magnetic activity of the Sun has been taken as a proxy whereas Rozanov et al. (2005) have ionization rates calculated for just one year, i.e. for a multiyear run, the IR values will repeat themselves.

The results for NO_y, ozone and temperature with the different IR values show that changes induced by just the LEE are several times weaker than the changes triggered by the LEE and HEE together. On the other hand, the pattern looks similar, which implies that the odd nitrogen produced through the low energetic electrons in the mesosphere is transported down with time. The fact that the NO_y produced at the model top thins out during the downward propagation explains why the changes are less pronounced when just using the LEE in the model run.

Beside the differences between the results analyzed above, one important difference among the GCRs, SPEs and EEPs is that they do not have the same time behaviour. All three are modulated by the solar cycle: GCRs approximately vary by a factor of 1.5 at the poles with lowest flux during solar maximum (Calisto et al., 2011). SPE frequency has its maximum in phase with the solar cycle (Manchester et al., 2005), whereas EEP's maximum is

shifted after the maximum (Turunen et al., 2008). SPEs and EEPs are highly variable and strong events with significant atmospheric effects occur at a rate of a few per year. In contrast, the flux of the GCRs and its energy spectrum shows only slow variations (apart from Forbush decreases).
Whereas all three particle types have an impact on ozone and radiative heating rates, and therefore on dynamics, the different spatial and time behaviour makes their distinct treatment in the models necessary. From our simulations it follows, that the effect of all three particles types have important effects in the atmosphere, and in addition, that also their combined effects can only be simulated by explicit treatment.
Therefore, it is important to have the GCRs, SPEs and EEPs parameterized in model runs. It is just as important to take into account that the model used for such calculations considers the coupling between physicochemical processes and large-scale dynamics.

8.2 Outlook

The output of the several model runs for the different energetic particles that have been performed in this thesis provide a unique data set for further exploration. Based on the presented results, several succeeding studies could be performed.
It would be interesting to perform a model run in which we enhance the intensity of the Galactic cosmic rays to investigate how the Earth's atmosphere reacts to this impact during times, when the magnetic poles are changing, i.e. when the magnetic shielding is negligible.
A similar topic has been analyzed by Winkler et al., (2008). They have used their 2-D model to look at the impact of solar protons during times when the magnetic field of the Earth is changing. They found that the flux of harmful ultraviolet radiation increases at the Earth's surface and that during large SPEs the ozone destruction is more pronounced and reaches deeper into the atmosphere than during ozone hole situations.
Because the GCRs are more energetic than the solar protons, the impact could be even more distinct than presented in Winkler et al., (2008).
This study could be extended including all types of particles to see how the atmosphere is changing during times of magnetic field line reversal.
An ongoing project is the comparison of the solar proton event from October/November 2003 with several models throughout the community (Funke

et al., 2011). The goal of this project is to find out how several models handle the same event with the same given ionization rates, so that one can say that, e.g. the 10% difference in the result is caused by the internal variabiliy of the model and not due to different IR.

To sum it up, the results and the parameterizations presented in this thesis provide the fundament for a variety of projects for the future and it can also be used as a work of reference when one is interested in the influence of the energetic particles on atmospheric chemistry and climate.

Appendix A
HEPPA intercomparison

In recent years, many new satellite instruments capable of polar measurements have been launched. This has given unique opportunities to study effects of particle precipitation in the middle atmosphere.

The HEPPA (High-Energy Particle Precipitation in the Atmosphere) community focuses on the observational as well as modelling studies of atmospheric and ionospheric changes caused by energetic particle precipitation, e.g. solar proton events, relativistic electron precipitation, and auroral electron precipitation.

When comparing the different results from the different models that are used in the HEPPA community, the question came up if the models or the parameterizations for the energetic particles are the main driver for the different results. In a second step, the model results are compared with the data obtained with the MIPAS instrument on the ENVISAT satellite.

To exclude the error of parameterization, the modelers from the HEPPA community used the same IR for the energetic particles. The model results obtained from this comparison are validated against the MIPAS data and are shown in this appendix.

The focus is laid on the SOCOL model which has been used in this thesis. The following pictures show the zonal mean from 70° to 90° N. The subsequent subchapters show data that are similar to Funke et al., 2011.

A.1 NOx comparison

The result for NO_x (= NO + NO_2) achieved with our model, which is visible in the lower panel in Figure A.1, is in a relatively good agreement with the satellite data (upper panel), even though the modeled result for NO_2 shows an underestimation compared with MIPAS data (not shown). This underestimation can be explained by a smaller decrease of ozone than shown with MIPAS (see Fig. A.3). Therefore, the production of NO_2 via $NO + O_3 \rightarrow NO_2 + O_2$ is less compared to the satellite data. NO, however, is in good agreement with the satellite data (not shown).

Most of the models that participated at the HEPPA comparison show similar restults for NO_x compared to SOCOL, i.e. most of them show an overestimation of odd nitrogen in the upper mesosphere and an underestimation in the lower parts of the atmosphere.

A.1 NOx comparison

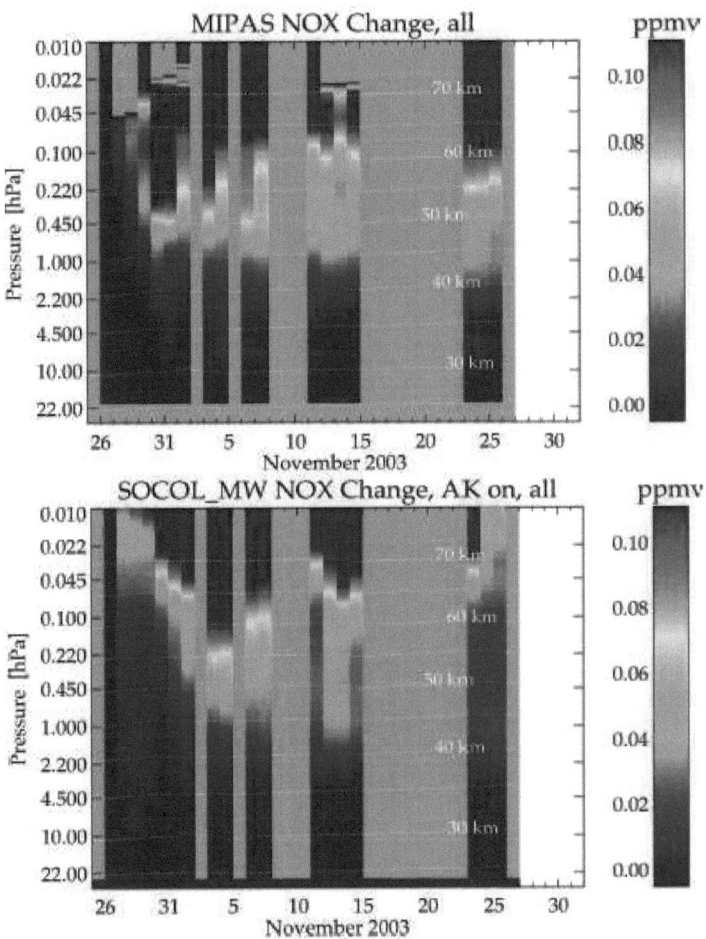

Figure A.1: *Changes in NOx induced by the solar proton event during October/November 2003 given in ppmv. Upper panel shows MIPAS satellite data, lower panel shows results obtained with SOCOL*

A.2 HNO$_3$ comparison

Comparing the lower panel in Figure A.2, which represents the results obtained with SOCOL, with the data obtained with MIPAS one sees that our model underestimates the production of HNO$_3$ (= OH + NO$_2$ + M $\rightarrow$ HNO$_3$ + M). This result is understandable, because there is also an underestimation in NO$_2$ (not shown) visible for SOCOL when comparing with the satellite data.

Evidently, all the models that participate in the HEPPA comparison have problems with HNO$_3$. None is able to represent the changes for HNO$_3$ similar to the satellite data. Keeping in mind that all the participants used the same ionization rates, one can conclude that the models are the source of the error.

A.2 HNO_3 comparison

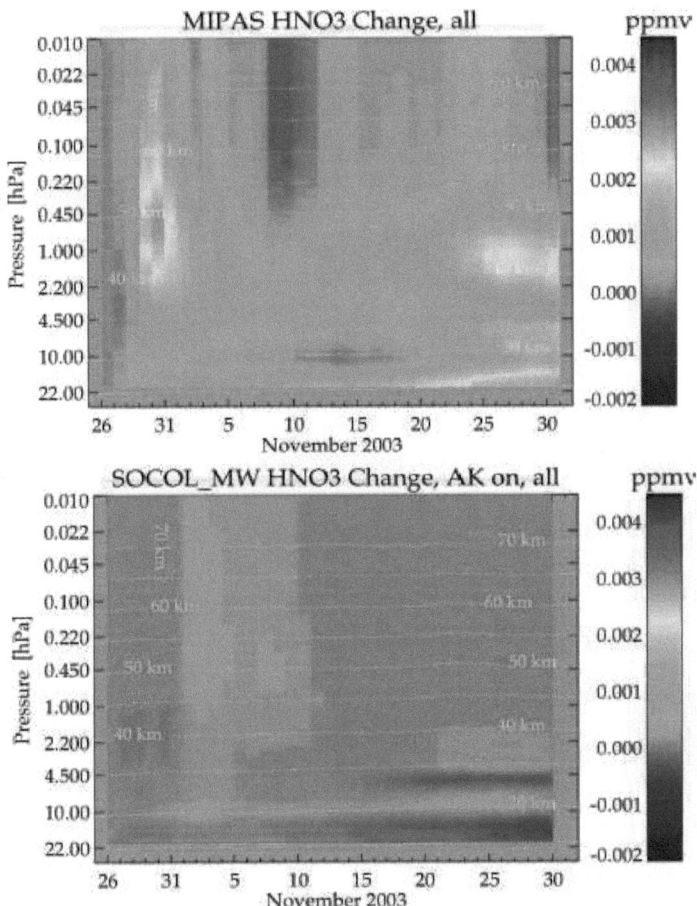

Figure A.2: *Changes in HNO3 induced by the solar proton event during October/November 2003 given in ppmv. Upper panel shows MIPAS satellite data, lower panel shows results obtained with SOCOL*

HEPPA intercomparison

A.3 Ozone comparison

Figure A.3 shows that the decrease of ozone through the Halloween storm modeled with SOCOL is in a relatively good agreement with the satellite data. A underestimation of about 40 % is depicted. The downward transport of the ozone depleted air is good visible in the lower panel which represents the result obtained with SOCOL.

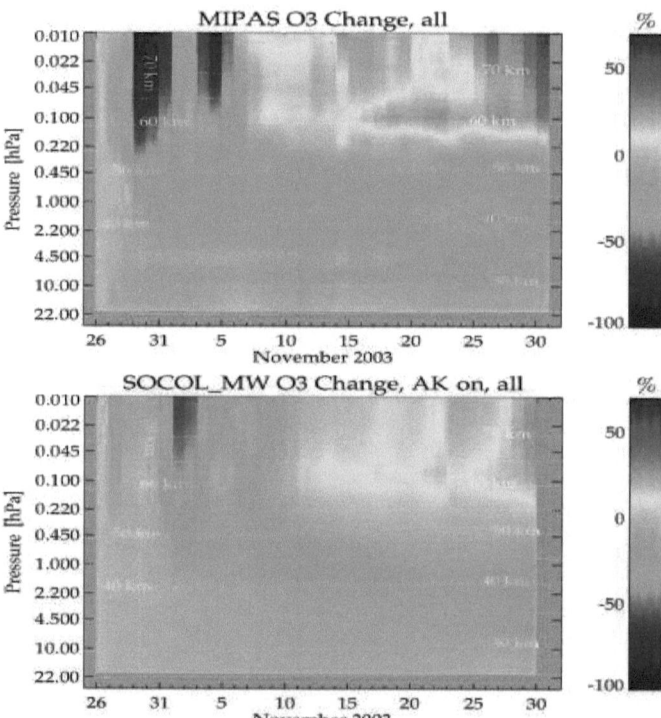

Figure A.3: *Changes in O3 induced by the solar proton event during October/November 2003 given in percent. Upper panel shows MIPAS satellite data, lower panel shows results obtained with SOCOL*

List of Figures

Comparing the other models with the satellite data one can see that most of them show an underestimation of about 20 to 40 % during the beginning of the solar proton event. The downward transport of the ozone depleted air works well throughout all the HEPPA models.

The outcome of the HEPPA comparison reveals that some models underestimate and others overestimate the effects of several chemical species induced by the SPE during October/Novmber 2003. It seems that the different resolutions, chemistry schemes and other internal variables of the models are a factor that needs to be taken into account.

List of Figures

2.1 Energy range for the different energetic particles. The GCRs are the most energetic, the SPEs and high energy electrons/relativistic electrons (HEEs/REPs) less, and low energy electrons (LEE) are the least energetic. 5

2.2 Altitude range where the Galactic cosmic rays (GCRs), solar protons (SPEs) and energetic electrons (EEPs) interact with the Earth's atmosphere. 6

2.3 Cyclicity of GCRs, taken from Kiel Neutron Monitor (http://cr0.izmiran.rssi.ru/kiel/main.htm) and Sunspots from 1958 to 2008. 8

2.4 Ionization rate (IR, number of ion pairs produced per air mass and time unit) for 90° geomagnetic latitude as computed by Usoskin, Kovaltsov & Mironova, (2010). Solid lines show IR during solar maximum and dashed lines during solar minimum. 9

2.5 Pathway of the Solar Protons to the Earth. (Taken from http://Sunclimate.gsfc.nasa.gov/projects/) 10

2.6 Earth with the Van Allen Belts and the trapped electrons. Taken from http://www.physics.sjsu.edu/becker/physics51/mag 13

2.7 Earth showing the auroral and sub-auroral regions. Taken from http://odin.gi.alaska.edu/FAQ/ 14

3.1 Main chemical reactions for the O_x family 18

3.2 Global map of the variations of total ozone, given in DU, measured by the TOMS instrument in 1990 (taken from Brasseur and Solomon, 2005.) 19

LIST OF FIGURES

3.3   Ozone number density in DU per kilometer averaged over a 10 year period. The data is based on the measurements of the Nimbus-7 SBUV instrument from 1980-1989. The black arrows in the figure represent the annual average of the Brewer-Dobson-Circulation in the stratosphere. Image taken from http://www.ccpo.odu.edu/~lizsmith/SEES/ozone/oz_class.htm   20

3.4   Polar vortex and its consequences. Taken from Climate & Society Lectures at Columbia University Department of Earth and Environmental Sciences: Climate and Society   28

4.1   The Earth's magnetic field is similar to that of a magnetic dipol. Clearly visible the entering of the lines close to the poles. Picture taken from ds9.ssl.berkeley.edu/themis/mission_magnetosphere.html   34

4.2   Overview of the different regions covered by Heaps parameterization.   35

4.3   Geomagnetic cutoff given in Gigavolts (GV). The closer to the equator, the larger the cutoff, i.e. only high energetic particles are able to penetrate deep into the equatorial atmosphere. . .   37

4.4   Ionization rates given in cm^{-3} s^{-1} starting end of October until the beginning of November provided from the SOLARIS website. The contour levels for the ionization rates are: 100, 200, 500, 800, 1000, 2000, 3000, 4000, 5000, 6000, 7000, 8000, 9000   39

4.5   Ionization rates given in cm^{-3} s^{-1} starting end of October until the beginning of November calculated with the ionization model AIMOS. The contour levels for the ionization rates are: 100, 200, 500, 800, 1000, 2000, 3000, 4000, 5000, 6000, 7000, 8000, 9000, 10'000   40

4.6   Ap index of the Sun for 2003. Clearly visible is the maximum during October/November 2003.   42

LIST OF FIGURES

5.1 Ionization rates (IR, number of ion pairs produced per air mass and time unit) for several geomagnetic latitudes as computed by the CRAC:CRII model (Usoskin, Kovaltsov & Mironova, 2010). Solid lines show ionization rates during solar maximum, the dashed lines during solar minimum. Note different scales on abscissas in dependence on geomagnetic latitude. 49

5.2 Annual mean effect of GCRs on zonal mean NO_x, $([NO_x]_{GCR}-[NO_x]_{control})/[NO_x]_{control}$, in percent ($[NO_x] = [NO] + [NO_2]$). Results are averaged from 1978-2002 (after allowing for a 2-year model spin-up) with appropriate accounting for solar minimum and maximum periods. Solid contours indicate positive, dotted contours negative changes. Hatched areas (enclosed by solid contours) indicate changes with at least 95 % statistical significance 55

5.3 Annual mean effect of GCRs on zonal mean HO_x, $([HO_x]_{GCR}-[HO_x]_{control})/[HO_x]_{control}$, in percent ($[HO_x] = [H] + [OH] + [HO_2]$). Results are averaged from 1978-2002 (after allowing for a 2-year model spin-up) with appropriate accounting for solar minimum and maximum periods. Solid contours indicate positive, dotted contours negative changes. Hatched areas (enclosed by solid contours) indicate statistically significant changes with at least 95 % (inner contours) or 80 % (outer contours). 56

5.4 Annual mean effect of GCRs on zonal mean HNO_3, $([HNO_3]_{GCR}-[HNO_3]_{control})/[HNO_3]_{control}$, in percent. Results are averaged from 1978-2002 (after allowing for a 2-year model spin-up) with appropriate accounting for solar minimum and maximum periods. Solid contours indicate positive, dotted contours negative changes. Hatched areas (enclosed by solid contours) indicate changes with at least 95 % statistical significance. 57

LIST OF FIGURES

5.5 Annual mean effect of GCRs on zonal mean ozone, ($[O_3]_{GCR}$-$[O_3]_{control}$)/$[O_3]_{control}$, given in percent. Results are averaged from 1978-2002 (after allowing for a 2-year model spin-up) with appropriate accounting for solar minimum and maximum periods. Solid contours indicate positive, dotted contours negative changes. Hatched areas (enclosed by solid contours) indicate changes with at least 95 % statistical significance. . . . 58

5.6 Monthly mean zonal mean effects of GCRs on ozone (O_3), temperature (T) and zonal wind (U) for the month of February. Upper panel: effect on O_3 given in percent (in steps of 0.5 %). Center panel: effect on T given in Kelvin (in steps of 0.25 K). Lower panel: effect on U given in m/s (in steps of 0.5 m/s). Hatched areas show 95 % statistical significance. 60

5.7 GCR-induced effects on ozone, ($[O_3]_{GCR}$-$[O_3]_{control}$)/$[O_3]_{control}$, given in percent. Left panel shows the annual mean averaged for 70° - 90° N and for 50° S (right). Red line: parameterization by Usoskin et al. (2010). Blue line: parameterization by Heaps (1976). Results are averaged from 1978-2002 (all seasons, after allowing for a 2-yr model spin-up) 62

5.8 GCR-induced effects on ozone, ($[O_3]_{GCR}$-$[O_3]_{control}$)/$[O_3]_{control}$, given in percent,f or 70° - 90° N. Upper panel: November and December; lower panel: February and March. Red lines: parameterization by Usoskin et al. (2010). Blue lines: parameterization by Heaps (1976). Thick solid lines: altitudes where the changes in ozone are significant at 95 % level for the respective parameterization. Results are averaged from 1978-2002. 63

5.9 Effect of GCRs on SAT, $[SAT]_{GCR}$-$[SAT]_{control}$, given in Kelvin for January monthly mean (left) and annual mean (right). Upper panels: using ionization rate modeled by Usoskin et al. (2010). Lower panels: using parameterization by Heaps (1978). Results are averaged from 1978-2002 (after allowing for a 2-yr model spin-up). Reddish colors: positive changes. Bluish colors: negative changes. Hatched areas (enclosed by thick solid lines) indicate changes with at least 95 % statistical significance. 64

LIST OF FIGURES

6.1 Time altitude profile of the ionization rates from the Carrington Event given in cm^{-3} s^{-1}. Contour levels: 0, 5, 20, 80, 200, 500, 800, 1'000, 4'000, 8'000, 12'000, 18'000, 25'000, 30'000. . . 76

6.2 Upper panel: Time altitude profile of NO$_x$ zonal mean from 70° - 90° N given in ppbv. Lower panel: same for 70° - 90° S. Colores indicate areas with at least 95 % statistical significance. Contour levels (ppbv): 10, 30, 50, 100, 200, 300, 400. 77

6.3 Upper panel: Time-altitude profile of HO$_x$ zonal mean change (in percent) for 70° - 90° N. Lower panel: same for southern hemisphere. Colors indicate areas with at least 95 % statistical significance. Contour levels for upper panel: -10, 0, 5, 20, 100, 500. Contour levels for lower panel: -10, 0, 20, 100, 400, 800, 1'200. 78

6.4 Upper panel: Time-altitude profile of ozone zonal mean change (in percent) from 70° - 90° N. Lower panel: same for southern hemisphere. Colors indicate areas with at least 95 % statistical significance. Contour levels (%): -100, -80, -40, -20, -10, 0, 2, 5, 10, 50. 79

6.5 Upper panel: Time-altitude profile of zonal mean temperature change (in Kelvin) for 70° - 90° N. Lower panel: same for the southern hemisphere. Colors indicate areas with at least 95 % statistical significance. Contour levels (K): -10, -4, -3, -2, -1, 0, 1, 2, 3, 5, 10. 81

6.6 Upper part: Polar stereographic projection of zonal wind changes at 25 hPa averaged from August to November (in m/s). Lower part: Zonal wind changes at 63 hPa averaged from August to November (in m/s). Left: northern hemisphere. Right: southern hemisphere. Hatched areas show 95 % statistical significance. 82

6.7 Left row: Polar stereographic projection of changes in total ozone (in DU) for the northern hemisphere from September to November resulting from the Carrington Event. Right row: Same for the southern hemisphere. Hatched areas show 95 % statistical significance. 84

LIST OF FIGURES

7.1 Annual mean changes of zonal mean NO_x, ($[NO_x]$exp-$[NO_x]$control), in ppb ($[NO_x] = [NO] + [NO_2]$). Results are averaged from 1962-2005. Solid contours indicate positive, dotted contours negative changes. Hatched areas (enclosed by solid contours) indicate changes with at least 95 % statistical significance 95

7.2 Annual mean changes of zonal mean NO_x, ($[NO_x]$exp-$[NO_x]$control), in ppb ($[NO_x] = [NO] + [NO_2]$). Results are averaged from 1962-2005. Solid contours indicate positive, dotted contours negative changes. Hatched areas (enclosed by solid contours) indicate changes with at least 95 % statistical significance 96

7.3 Annual mean changes of zonal mean HO_x, ($[HO_x]$exp-$[HO_x]$control)/$[HO_x]$control, in percent ($[HO_x] = [H] + [OH] + [HO_2]$). Results are averaged from 1962-2005. Hatched areas (enclosed by solid contours) indicate statistically significant changes with at least 95 %. 97

7.4 Annual mean changes of zonal mean HNO_3, ($[HNO_3]$exp-$[HNO_3]$control), in ppt from 1000 hPa up to 200 hPa. Results are averaged from 1962-2005. Solid contours indicate positive, dotted contours negative changes. Hatched areas (enclosed by solid contours) indicate changes with at least 95 % statistical significance 98

7.5 Annual mean changes of zonal mean HNO_3, ($[HNO_3]$exp-$[HNO_3]$control), in ppt from 200 hPa up to 1 hPa. Results are averaged from 1962-2005. Solid contours indicate positive, dotted contours negative changes. Hatched areas (enclosed by solid contours) indicate changes with at least 95 % statistical significance 99

7.6 Annual mean changes of zonal mean HNO_3, ($[HNO_3]$exp-$[HNO_3]$control), in ppt from 1 hPa up to 0.01 hPa. Results are averaged from 1962-2005. Solid contours indicate positive, dotted contours negative changes. Hatched areas (enclosed by solid contours) indicate changes with at least 95 % statistical significance 100

LIST OF FIGURES

7.7 Annual mean changes of zonal mean ozone, ($[O_3]$exp-$[O_3]$control)/$[O_3]$control, given in percent. Results are averaged from 1962-2005. Solid contours indicate positive, dotted contours negative changes. Hatched areas (enclosed by solid contours) indicate changes with at least 95 % statistical significance. 101

7.8 Upper panel: Monthly mean changes for zonal mean ozone for November given in percent. Lower panel: Monthly mean changes of zonal mean zonal wind for November given in m/s. Hatched areas (enclosed by solid contours) indicate changes with at least 95 % statistical significance. 102

7.9 Annual mean changes for total ozone given in DU due to the GCRs, SPEs and LEEs. Results are averaged from 1962 to 2005. Left panel represents the NH whereas the SH is on the right side. Hatched areas (enclosed by solid contours) indicate changes with at least 95 % statistical significance. 103

A.1 Changes in NOx induced by the solar proton event during October/November 2003 given in ppmv. Upper panel shows MIPAS satellite data, lower panel shows results obtained with SOCOL 111

A.2 Changes in HNO3 induced by the solar proton event during October/November 2003 given in ppmv. Upper panel shows MIPAS satellite data, lower panel shows results obtained with SOCOL 113

A.3 Changes in O3 induced by the solar proton event during October/November 2003 given in percent. Upper panel shows MIPAS satellite data, lower panel shows results obtained with SOCOL 114

LIST OF FIGURES

List of Tables

2.1 Altitude at which electrons of a given energy produce the maximum ionization rate for several geomagnetic latitudes (λ_m). The heights at which the ionization rates have decreased to 10 % below and above the maximum are also given in this table. (Taken from M. H. Rees, 1964) 15

4.1 Overview of the Parameters A and B. 36

4.2 Overview of the Parameters C and D. 36

4.3 Overview of the model experiments. 46

LIST OF TABLES

Bibliography

[1] Aikin, A.C., Energetic particle-induced enhancements of stratospheric nitric acid, *Geophys. Res. Lett.*, *21*, 859-862, 1994

[2] Baumgaertner, A.J.G., Joeckel, P. and Bruehl, C., Energetic particle precipitation in ECHAM5/MESSy1 - Part 1: Downward transport of upper atmospheric NO_x produced by low energy electrons, *Atmos. Chem. Phys.*, *9*, 2729-2740, 2009

[3] Barr, R., Jones, D.L. and Rodger, C.J., ELF and VLF radio waves, *J. Atmos. Terr. Phys.*, *62*, 1689-1718, 2000

[4] Bazilevskaya, G. et al., Physics of auroral phenomena, *paper presented at the XXV annual seminar, Polar Geophys. Inst.*, Apatity, Russia, 2002

[5] Bazilevskaya, G.A., Usoskin, I.G., Flueckiger, E.O., Harrison, R.G., Desorgher, L., Buetikofer, R., Krainev, M.B., Makhmutov, V.S., Stozhkov, Y.I., Svirzhevskaya, A.K., Svirzhevsky, N.S. and Kovaltsov, G.A., Cosmic Ray Induced Ion Production in the Atmosphere, *Space Sci. Rev.*, *137*, 149, 2008

[6] Bhabha, H.J. and Heitler, W., The passage of fast electrons and the theory of Cosmic showers, *Proc. Roy. Soc. Lond. Math. Phys. Sci.*, *898*, 432-458, 1937

[7] Brasseur, G. and Solomon, S., Aeronomy of the Middle Atmosphere, *Springer, Dordrecht, 3rd edition*, 2005

[8] Brewer, A.W., Evidence for a world circulation provided by measurements of helium and water vapor distribution in the stratosphere, *Quart. J. Roy. Meteorol. Soc.*, *75*, 351-363, 1949

[9] Bridge, H.S., Peyrou, C., Rossi, B. and Saffort, R., Cloud-chamber observations of the heavy charged unstable particles in Cosmic rays, *Phys. Rev.*, *90*, 921-933, 1953

Bibliography

[10] Calisto, M., Usoskin, I., Rozanov, E. and Peter, T, Influence of galactic cosmic rays on atmospheric composition and temperature, *Atmos. Chem. Phys. Discuss.*, *11*, 653-679, 2011

[11] Callis, L. Natarajan, M., Lambeth, J.D. and Baker, D.N., Solar atmospheric coupling by electrons (SOLACE): 2.Calculated stratospheric effects of precipitating electrons, 1979-1988, *J. Geophys. Res.*, *103*, 28421-28438, 1998

[12] Carslaw, K.S., Luo, B.P. and Peter, T., An analytic expression for the composition of aqueous HNO_3-H_2SO_4 stratospheric aerosols including gas phase removal of HNO_3, *Geophys. Res. Lett.*, *22*, 1877-1880, 1995

[13] Chubachi, S., Preliminiary result of ozone observations at Syowa Station from February, 1982, to January, 1983, *Mem. Natl. Inst. Polar Res. Jap., Spec Issue 34*, 13, 1984

[14] Crutzen, P. J., The influence of nitrogen oxides on the atmospheric ozone content, *Q. J. Roy. Meteorol. Soc.*, *96*, 320-325, 1970

[15] Dobson, G.M.G., Origin and distribution of polyatomic molecules in the atmosphere, *Proc. Roy. Soc. Lond. A, 236*, 187-193, 1956

[16] Egorova, T., Rozanov, E., Zubov, V. and Karol, I.L., Model for Investigating Ozone Trends (MEZON), Izvestiya, *Atmospheric and Oceanic Physics, 39*, 277-292, 2003

[17] Egorova, T., Rozanov, E., Ozolin, Y., Shapiro, A., Calisto, M., Peter, Th. and Schmutz, W., The atmospheric effects of October 2003 solar proton event simulated with the chemistry-climate model SOCOL using complete and parameterized ion chemistry, *J. Atmos. Sol. Terr. Phys., 73*, 356-365, 2011

[18] Elkington, S. R., Hudson, M. K. and Chan, A. A., Enhanced Radial Diffusion of Outer Zone Electrons in an Asymmetric Geomagnetic Field, *Spring Meeting 2001. American Geophysical Union*

[19] European Space Agency, Envisat, MIPAS: An Instrument for Atmospheric Chemistry and Climate Research, *ESA Publ. Div.*, ESTEC, Noordwijk, Netherlands, 2000

Bibliography

[20] Farman, J.C., Gardiner, B.G. and Shanklin, J.D., Large losses of total ozone in Antarctica reveal seasonal ClO_x/NO_x interaction, *Nature, 315*, 207-210, 1985

[21] Funke, B., Baumgaertner, A., Calisto, M., Egorova, T., Jackman, C.H., Kieser, J., Krivolutsky, A., Lopez, M., Marsh, D.R., Reddmann, T., Rozanov, E., Salmi, S.-M., Sinnhuber, M., Stiller, G.P., Verronen, P.T., Versick, S., von Clarmann, T., Vyushkova, T.Y., Wieters, N. and Wissing, J.M., Composition changes after the Halloween solar proton event: the High-Energy Particle Precipitation in the Atmosphere (HEPPA)model versus MIPAS data intercomparison study, *submitted to Atmos. Chem. Phys.*, 2010

[22] Hanson, D.R. and Ravishankara, A.R., Heterogeneous chemistry of bromine species in sulfuric acid under stratospheric conditions, *Geophys. Res. Lett., 22*, 385-388, 1995

[23] Hanson, D., Ravishankara, A. and Lovejoy, E., Reaction of $BrONO_2$ with H_2O on submicron sulfuric acid aerosol and the implications for the lower stratosphere, *J. Geophys. Res., 101*, 9063-9069, 1996

[24] Heaps, M.G., Parametrization of the cosmic ray ion-pair production rate above 18 km, *Planet. Space Sci., 26*, 513-517, 1978

[25] Holton, J. R., An Introduction to Dynamic Meteorology, *4th ed., Elsevier Academic Press, ISBN: 0-12-354016-X*, 2004

[26] Hoppel, K.W., Bowman, K.P. and Bevilacqua, R.M., Northern hemisphere summer ozone variability observed by POAM II, *Geophys. Res. Lett., 26*, 827-830, 1999

[27] Horne, R. B.; Thorne, R. M. et al., Wave acceleration of electrons in the Van Allen radiation belts, *Nature, 437*, 227-230, 2005

[28] Hudson, M.K., Kress, B.T., Mueller, H.-R., Zastrow, J.A. and Blake, J.B., Relationship of the Van Allen radiation belts to solar wind drivers, *J. Atmos. Sol. Terr. Phys., 70*, 708-729, 2008

[29] Krivolutsky, A., Bazilevskaya, G., Vyushkova, T. and Knyazeva, G., Influence of cosmic rays on chemical composition of the atmosphere: data analysis and photochemical modeling, *Phys. and Chem. of the Earth, 27*, 471-476, 2002

Bibliography

[30] Lal, D., Jull, A.J.T., Pollard, D. and Vacher, L., Evidence for large century time-scale changes in solar activity in the past 32 Kyr, based on in-situ cosmogenic ^{14}C in ice at Summit, Greenland, *Earth Planet. Sci. Lett.*, *234*, 335-349, 2005

[31] Lopez, R.E. and Baker, D.N., Evidence for particle acceleration during magnetospheric substorms, *Astrophys. J. Suppl.*, *90*, 531-539, 1994

[32] Legrand, M.R., Stordal, F., Isaksen, I.S.A. and Rognerud B., A model study of the stratospheric budget of odd nitrogen, including effecs of solar cycle variations, *Tellus*, *41B*, 413-426, 1988

[33] Manchester IV, W.B., Gombosi, T.I., De Zeeuw, D.L., Sokolov, I.V., Roussev, I.I., Powell, K.G., Kota, J., Toth, G. and Zurbuchen, T.H., Coronal mass ejection shock and sheath structures relevant to particle acceleration, *Astrophys. J.*, *622*, 1225-1239, 2005

[34] Manzini, E., McFarlane, N.A. and McLandress, C., Impact of the Doppler spread parameterization on the simulation of the middle atmosphere circulation using the MA/ECHAM4 general circulation model, *J. Geophys. Res. Atmos.*, *102*, D22, 25751-25762, 1997.

[35] McCormick, M.P., Steel, H.M., Hamill, P., Chu, W.P. and Swissler, T.J., Polar stratsopheric cloud sightings by SAM II, *J. Atmos. Sci.*, *39*, 1387-1397, 1982

[36] McElroy, M.B., Salawitch, R.J. and Minschwaner, K., The changing stratosphere, *Planet. Space Sci.*, *40*, 373-401, 1992

[37] Molina, L.T. and Molina, M.J., Production of Cl_2O_2 from the self-reaction of ClO radical, *J. Phys. Chem.*, *91*, 433-436, 1987

[38] Molina, J.S. and Rowland, F.S., Stratospheric sink for chlorofluoromethanes: Chlorine atom-catalyzed destruction of ozone, *Nature*, *249*, 810-812, 1974

[39] Neher, H.V., Low-energy primary cosmic-ray particles in 1954, *Phys. Rev.*, *103*, 228-236, 1956

[40] Norval, M., Cullen, A.P., DeGruijl, F.R., Longstreth, J., Takizawa, Y., Lucas, R.M., Noonan, F.P. and Van der Leun, J.C., The effects on human health from stratospheric ozone depletion and its interactions with climate change, *Photochem. Photobiol. Sci.*, *6*, 232-251, 2007

Bibliography

[41] Porter, H.S., Jackman, C.H. and Green, A.E.S., Efficiencies for production of atomic nitrogen and oxygen by relativistic proton impact in air, *J. Chem. Phys.*, *65*, 1, 1976

[42] Prather, M.J., Numerical Advection by conservation of 2nd-order moments, *J. Geophys. Res. Atmos.*, *91, D6*, 6671-6681, 1986

[43] Rees, M.H., Note on the penetration of energetic electrons into the Earth's atmosphere, *Planet. Space Sci.*, *12*, 722-725, 1964

[44] Rodger, C.J., Verronen, P.T., Clilverd, M.A., Seppaelae, A. and Turunen, E., Atmospheric impact of the Carrington event solar protons, *J. Geophys. Res.*, *113*, doi:10.1029/2008JD010702, 2008

[45] Rossi, B.B. and Staub, H.H., Ionization chambers and counters: Experimental Techniques, *McGray-Hill Book Company, Inc*, First edition, 1949

[46] Rozanov, E., Schlesinger, M.E., Zubov, V., Yang, F. and N. Andronova, G., The UIUC three-dimensional stratospheric chemical transport model: Description and evaluation of the simulated source gases and ozone, *J. Geophys. Res.*, *104*, 11755-11781, 1999

[47] Rozanov. E., Callis, L., Schlesinger, M., Yang, F. Andronova, N. and Zubov, V., Atmospheric response to NO_y source due to energetic electrons precipitation, *Geophys. Res. Lett.*, *32*, doi:10.1029/2005GL023041, 2005

[48] Schraner, M., Rozanov, E., Schnadt-Poberaj, C., Kenzelmann, P., Fischer, A., Zubov, V., Luo, B.P., Hoyle, C., Egorova, T., Fueglistaler, S., Broennimann, S., Schmutz W. and Peter, T., Chemistry climate model SOCOL: version 2.0 with improved transport and chemistry/ microphysics schemes, *Atmos. Chem. Phys.*, *8, 19*, 5957-5974, 2008

[49] Shindell, D.T., Schmidt, G.A., Miller, R.L. and Rind, D., Northern Hemisphere winter climate response to greenhouse gas, ozone, solar, and volcanic forcing, *J. Geophys. Res.*, *106*, 7193-7210, 2001

[50] Shprits, Y.Y. and Thorne, R.M., Time dependent radial diffusion modeling of relativistic electrons with realistic loss rates, *Geophys. Res. Lett.*, *31*, doi:10.1029/2004GL019591, 2004

Bibliography

[51] Shprits, Y.Y., Thorne, R.M., Horne, R.B., Glauert, S.A., Cartwright, M., Russel, C.T., Baker, D.N. and Kanekal, S.G., Acceleration mechanism resposible for the formation of new radiation belt during the 2003 Halloween solar storm, *Geophys. Res. Lett.*, *33*, doi:10.1029/2005GL024256, 2006

[52] Shprits, Y.Y., Thorne, R.M., Friedel, R., Reeves, G.D., Fennell, J., Baker, D.N. and Kanekal, S.G., Outward radial diffusion driven by losses at magnetopause, *J. Geophys. Res.*, *111*, doi:10.1029/2006JA011657, 2006

[53] Solomon, S., Rusch, D.W., Gerard, J.-C., Reidt, G.C. and Crutzen, P.J., The effect of particle precipitation events on the neutral and ion chemistry of the middle atmosphere: II. Odd Hydrogen, *Planet. Space Sci.*, *29*, 8, 885-892, 1981

[54] Solomon, S., Garcia, R., Rowland, R.R. and Wuebbles, D.J., On the depletion of Antarctic ozone, *Naure*, *321*, 755-758, 1986

[55] Smart, D.F., Shea, M.A. and McCracken, K.G., The Carrington event: Possible solar proton intensity-time profile, *Adv. Space Res.*, *38*, 215-225, 2005

[56] Tascione, T.F., Introduction to the Space Environment, 2nd. Ed., *Kreiger Publishing CO.*, Malabar, Florida USA, 1994

[57] Thompson, D.W.J. and Wallace, J.M., The Arctic Oscillation signature in the wintertime geopotential height and temperature fields, *Geophys. Res. Lett.*, *25*, 9, 1297-1300, 1998

[58] Thorne, R.M., O'Brien, T.P., Shprits, Y.Y., Summers, D. and Horne, R.B., Timescale for MeV electron microburst loss during geomagnetic storms, *J. Geophys. Res.*, *110*, doi:10.1029/2004JA010882, 2005

[59] Turunen, E., Verronen, P.T., Seppaelae, A., Rodger, C.J., Clilverd, M.A., Tamminen, J., Enell, C.-F. and Ulich, T., Impact of different energies of precipitating particles on NO_x generation in the middle and upper atmosphere during geomagnetic storms, *J. Atmos. Sol. Terr. Phys.*, *71*, 1176-1189, 2009

[60] Usoskin I.G., Gladysheva, O.G. and Kovaltsov, G.A., Cosmic ray-induced ionization in the atmosphere: spatial and temporal changes, *J. Atmos. Sol. Terr. Phys.*, *66*, 1791, 2004

Bibliography

[61] Usoskin I.G., Alanko-Huotari K., Kovaltsov G.A. and Mursula, K., Heliospheric modulation of cosmic rays: Monthly reconstruction for 1951-2004, *J. Geophys. Res., 110*, A12108, 2005

[62] Usoskin, I.G. and Kovaltsov, G.A. Cosmic ray induced ionization in the atmosphere: Full modeling and practical applications, *J. Geophys. Res., 111, D21206*, doi:10.1029/2006JD007150, 2006

[63] Usoskin, I.G., Desorgher, L., Velinov, P., Storini, M., Flueckiger, E.O., Buetikofer, R. and Kovaltsov, G.A., Ionization of the Earth's atmosphere by solar and Galactic cosmic rays, *Acta Geophysica, 57*, 88-101, 2009

[64] Usoskin, I.G., Kovaltsov, G.A. and Mironova, I.A., Cosmic ray induced ionization model CRAC:CRII : An extension to the upper atmosphere, *J. Geophys. Res.*, doi:10.1029/2009JD013142, 2010

[65] Verronen, P.T., Seppaelae, A., Clilverd, M.A., Rodger, C.J., Kyroelae, E., Enell, C.F., Ulich, Th. and Turunen, E., Diurnal variation of ozone depletion during the October-November 2003 solar proton events, *J. Geophys. Res., 110*, doi:10.1029/2004JA010932, 2005

[66] Wagner, R.E. and Bowman, K.P., Wavebreaking and mixing in the Northern hemisphere summer stratosphere, *J. Geophys. Res., 105*, 24799-24807, 2000

[67] Weibull, W., A statistical distribution function of wide applicability, *J. Appl. Mech., 18*, 293-297

[68] Weeks, L. H., Cuikay, R. S. and Corbin, J. R., Ozone Measurements in the Mesosphere during the Solar Proton event of 2 November 1969, *J. Atmos. Sci., 29*, 1138-1142, 1972

[69] Williamson, D.L. and Rasch, P.J., Two-Dimensional Semi-Lagrangian Transport with shape-preserving interpolation, *Monthly Weather Review, 117, 1*, 102-129, 1989

[70] Winkler, H., Sinnhuber, M., Notholt, J., Kallenrode, M.-B., Steinhilber, F., Vogt, J., Zieger, B., Glassmeier, K.-H. and Stadelmann A., Modeling impacts of geomagnetic field variations on middle atmospheric ozone responses to solar proton events on long timescales, *J. Geophys. Res., 113*, doi:10.1029/2007JD008574, 2008

[71] Wissing, J.M. and Kallenrode, M.-B., Atmospheric Ionization Module Osnabrueck (AIMOS): A 3-D model to determine atmospheric ionization by energetic charged particles from different populations, *J. Geophys. Res.*, *114*, doi:10.1029/2008JA013884, 2009

[72] Yoshida, S., Polar Cap Absorption on VLF emissions, *Planet. Space Sci.*, *13*, 1165-1170, 1965

Die VDM Verlagsservicegesellschaft sucht für wissenschaftliche Verlage abgeschlossene und herausragende

Dissertationen, Habilitationen, Diplomarbeiten, Master Theses, Magisterarbeiten usw.

für die kostenlose Publikation als Fachbuch.

Sie verfügen über eine Arbeit, die hohen inhaltlichen und formalen Ansprüchen genügt, und haben Interesse an einer honorarvergüteten Publikation?

Dann senden Sie bitte erste Informationen über sich und Ihre Arbeit per Email an *info@vdm-vsg.de*.

Sie erhalten kurzfristig unser Feedback!

VDM Verlagsservicegesellschaft mbH
Dudweiler Landstr. 99   Telefon +49 681 3720 174
D - 66123 Saarbrücken   Fax   +49 681 3720 1749
www.vdm-vsg.de

Die VDM Verlagsservicegesellschaft mbH vertritt

Printed by Books on Demand GmbH, Norderstedt / Germany